Applications of Blockchain and Computational Intelligence in Environmental Sustainability

The book explores the complex correlation between blockchain technology and sustainability, demonstrating the potential of cutting-edge computational intelligence methods to address critical environmental and societal challenges. It provides a comprehensive analysis of the most recent developments in research, innovative approaches to design, and real-world implementations, establishing a strategic plan for the incorporation of blockchain technology into environmentally friendly solutions.

Features:

- Focuses on the intersection of blockchain technology, computational intelligence, and sustainability.
- Covers international regulatory landscapes and ethical considerations.
- Emphasizes real-world applications such as supply chain management systems and smart energy networks.
- Offers concrete examples of how these technologies contribute to sustainability.
- Provides insight into how new technologies are transforming health informatics.

This book is aimed at graduate students and researchers in computer engineering and environmental sustainability.

Computational Intelligence Techniques

Series Editor: Vishal Jain

The objective of this series is to provide researchers a platform to present state of the art innovations, research, and design and implement methodological and algorithmic solutions to data processing problems, designing and analyzing evolving trends in health informatics and computer-aided diagnosis. This series provides support and aid to researchers involved in designing decision support systems that will permit societal acceptance of ambient intelligence. The overall goal of this series is to present the latest snapshot of ongoing research as well as to shed further light on future directions in this space. The series presents novel technical studies as well as position and vision papers comprising hypothetical/speculative scenarios. The book series seeks to compile all aspects of computational intelligence techniques from fundamental principles to current advanced concepts. For this series, we invite researchers, academicians and professionals to contribute, expressing their ideas and research in the application of intelligent techniques to the field of engineering in handbook, reference, or monograph volumes.

Convergence of IoT, Blockchain and Computational Intelligence in Smart Cities
Edited by Rajendra Kumar, Vishal Jain, Leong Wai Yie, and Sunantha Prime Teyarachakul

Intelligent Techniques for Cyber-Physical Systems
Edited by Mohammad Sajid, Anil Kumar, Jagendra Singh, Osamah Ibrahim Khalaf, and Mukesh Prasad

Big Data Computing
Advances in Technologies, Methodologies, and Applications
Edited by Tanvir Habib Sardar and Bishwajeet Kumar Pandey

Software Defined Network Frameworks
Security Issues and Use Cases
Edited by Mandeep Kaur, Vishal Jain, Parma Nand and Nitin Rakesh

Machine Learning Hybridization and Optimization for Intelligent Applications
Edited by Tanvir Habib Sardar and Bishwajeet Kumar Pandey

Information Technology Ethics
Hamed Taherdoost

Applications of Blockchain and Computational Intelligence in Environmental Sustainability
Edited by Hamed Taherdoost, Mohsen Saeedi and Aydin Shishegaran

For more information about this series, please visit: www.routledge.com/Computational-Intelligence-Techniques/book-series/CIT

Applications of Blockchain and Computational Intelligence in Environmental Sustainability

Edited by Hamed Taherdoost, Mohsen Saeedi
and Aydin Shishegaran

CRC Press
Taylor & Francis Group
Boca Raton London New York

CRC Press is an imprint of the
Taylor & Francis Group, an **informa** business

Designed cover image: Shutterstock

First edition published 2025
by CRC Press
2385 NW Executive Center Drive, Suite 320, Boca Raton FL 33431

and by CRC Press
4 Park Square, Milton Park, Abingdon, Oxon, OX14 4RN

CRC Press is an imprint of Taylor & Francis Group, LLC

ISBN: 978-1-032-81513-8 (hbk)
ISBN: 978-1-041-00441-7 (pbk)
ISBN: 978-1-003-60986-5 (ebk)

DOI: 10.1201/9781003609865

Typeset in Times
by Apex CoVantage, LLC

Contents

Preface

In a world increasingly driven by technology and a pressing need for sustainable solutions, the intersection of blockchain technology and sustainability represents a beacon of hope and innovation. This book delves into this dynamic interplay, offering a comprehensive exploration of how blockchain technology can be and is being leveraged to foster sustainable practices across various sectors.

PURPOSE AND SCOPE

The primary aim of this book is to provide readers with an in-depth understanding of the potential and practical applications of blockchain technology in promoting sustainability. Through detailed chapters, we explore a range of topics from the foundational principles of blockchain and smart contracts to their applications in renewable energy, agriculture, and circular economies. Each chapter is meticulously structured to cover theoretical aspects, real-world applications, challenges, and future outlooks.

STRUCTURE OF THE BOOK

The book is organized into nine chapters, each focusing on a distinct aspect of blockchain technology and its relevance to sustainability:

1. **Introduction to Blockchain and Sustainability**: This chapter sets the stage with an overview of blockchain technology, its key features, and its potential to enhance sustainability through transparency, traceability, and security.
2. **Blockchain in Environmental Conservation:** This chapter discusses how blockchain technology, known for underpinning cryptocurrencies like Bitcoin, is revolutionizing environmental conservation. By providing decentralized, transparent ledgers, blockchain enhances sustainable practices, carbon trading, resource management, wildlife protection, and renewable energy initiatives. As the technology develops, its applications in environmental conservation and broader industrial and societal changes are expected to grow, offering innovative solutions to significant environmental challenges.
3. **Smart Contracts and Sustainable Business Models**: This chapter examines how smart contracts can revolutionize business models by embedding sustainability into their core operations, highlighting the challenges and opportunities for innovation.
4. **Blockchain Integration in Renewable Energy**: This chapter explores the applications of blockchain in the renewable energy sector, emphasizing the benefits and hurdles in integrating these technologies to foster a sustainable energy future.

5. **Sustainable Agriculture and Food Supply Chains**: This chapter explores the critical issues facing food supply chains and how blockchain can address these challenges, from resource optimization to minimizing emissions.
6. **Blockchain Applications in Sustainable Agriculture and Food Systems**: Building on the previous chapter, this chapter provides case studies from various countries, showcasing the practical use of blockchain in enhancing agricultural sustainability.
7. **Blockchain and Circular Economy**: This chapter discusses the principles of a circular economy and the transformative role blockchain can play in facilitating sustainable economic practices.
8. **Sustainability Reporting and Transparency**: The focus here is on how blockchain can enhance transparency and accountability in sustainability reporting, offering a foundation for future advancements in this area.
9. **Blockchain for Carbon Credits and Emissions Reduction**: This chapter addresses the integration of blockchain in the carbon credit market, exploring how this technology can streamline carbon trading and support global emissions reduction efforts.

AUDIENCE

This book is intended for a diverse audience including scholars, industry professionals, policymakers, and anyone interested in the potential of blockchain technology to drive sustainable development. Whether you are a technologist seeking to understand sustainability applications, a business leader looking to integrate blockchain into sustainable practices, or a researcher exploring new frontiers, this book offers valuable insights and practical knowledge.

ACKNOWLEDGMENTS

The creation of this book has been a collaborative effort, drawing on the expertise and contributions of numerous individuals. We extend our gratitude to the researchers, practitioners, and experts who have provided their insights and to the academic and professional communities that continue to push the boundaries of what is possible in the realms of blockchain and sustainability.

We hope this book serves as a valuable resource and inspiration for all who read it, fostering a deeper understanding and appreciation of the profound impact that blockchain technology can have on our journey toward a more sustainable future.

About the Editors

Dr. Hamed Taherdoost is an award-winning leader in research and development, recognized for his significant contributions across industry and academia. He is the founder of Hamta Business Corporation, Associate Professor and Chair of RSAC at University Canada West, and Director of R&D at Q Minded. With over 20 years of experience, Dr. Taherdoost has worked in global companies based in Cyprus, the UK, Malta, Iran, Malaysia, and Canada, and has been instrumental in various industry projects across healthcare, transportation, residential, oil and gas, and IT sectors. Additionally, he has served as a technical and technology consultant for numerous companies, providing expert guidance and mentorship.

In academia, Dr. Hamed Taherdoost has held teaching roles across Southeast Asia, the Middle East, Europe, and North America since 2009. Actively engaged in academic publishing and advisory roles, he serves on the editorial boards of prestigious journals from leading publishers. His academic leadership extends to organizing and chairing numerous workshops and conferences, as well as contributing to scientific and technical committees for over 300 international conferences worldwide.

Dr. Taherdoost has a prolific publishing record, with over 300 scientific articles in top-tier journals and conference proceedings. His work is widely recognized, with a strong citation impact reflecting in a high h-index, and his contributions span book chapters, edited volumes, and authored books focusing on technology and research methodology. His research accomplishments have earned him a place among the top 10 SSRN Business Authors from 2022 to 2024, as well as repeated listings on the Stanford-Elsevier compilation of the world's top 2% scientists from 2021 to 2024.

He currently serves as Editor for the Routledge (Taylor & Francis Group) Book Series *Mastering Academic Excellence: Research, Teaching, Learning, and Publishing* and is Editor of several prominent journals, including the *International Journal of Information Technology Project Management* (IGI), *EAI Endorsed Transactions on Scalable Information Systems, Journal of Blockchain* (MDPI), and *International Journal of Data Mining, Modelling, and Management* (Inderscience). Dr. Taherdoost is also an Associate Editor of *Frontiers in Research Metrics and Analytics*. Dr. Taherdoost is a GUS Fellow at the GUS Institute | Global University Systems and holds senior memberships in IEEE and Working Group Member of the International Federation for Information Processing (IFIP) TC 11.

Dr. Mohsen Saeedi is Professor of Sustainability at the University of Canada West. With over 20 years of experience in academia, he specializes in sustainability, environmental management, and technology, having contributed to both research and curriculum development in Canada and Iran. He has served in prominent roles, including Dean of the School of Civil Engineering at the Iran University of Science and Technology. His research spans sustainable development and environmental pollution, environmental management, and management systems, with numerous publications in peer-reviewed journals. He also serves as an associate editor for international sustainability-focused journals. As an editor of this book, he brings his expertise to the intersection of blockchain technology and sustainable practices.

Aydin Shishegaran is Lecturer at IU International University of Applied Sciences and another institute in Germany. He got his first PhD in 2022 from Iran University of Science and Technology and his second PhD accepted in 2024 at Bauhaus University Weimar. However, he has published over pre-reviewed papers in high-ranked journals and has 798 citations and has also reviewed 265 manuscripts for high-ranked journals. He is an editor in *Journal of Mechatronics and Artificial Intelligence in Engineering* since 2020. Furthermore, in 2021, he is a guest editor in Sustainability at MDPI. In 2020, he published a new MCDA method for sustainable evaluation. In 2023, 2023, and 2024, he published four new ML methods in high-ranked journal, in which one of them is the most cited paper in computers and structures.

Contributors

Ahmad Abu-Alkheil
Gisma University of Applied Sciences
Potsdam, Germany

Reza Ahmadi
Gisma University of Applied Sciences
Potsdam, Germany

Mohammad Amin Borghei
Gisma University of Applied Sciences
Potsdam, Germany

Vahab Esfandani
Gisma University of Applied Sciences
Potsdam, Germany

Seyed Yasin Jamali
Bauhaus Universität Weimar
Weimar, Germany

Peter Konhäusner
Gisma University of Applied Sciences
Potsdam, Germany

Mitra Madanchian
Hamta Group, Hamta Business
   Corporation,
Vancouver, Canada

Department of Arts, Communications
   and Social Sciences
University Canada West
Vancouver, Canada

Sara ravan Ramzani
Gisma University of Applied Sciences
Potsdam, Germany

Mojtaba Rezaie
Department of Civil Engineering
Islamic Azad University
Iran

Mohsen Saeedi
University Canada West
Vancouver, Canada

Aydin Shishegaran
IU International University of Applied
   Sciences
Dresden, Germany

Jack Smith
University Canada West
Vancouver, Canada

Hamed Taherdoost
Q Minded, Quark Minded
   Technology Inc.,
Vancouver, Canada
GUS Institute I Global University
   Systems
London, UK
University Canada West
Vancouver, Canada

1 Introduction to Blockchain and Sustainability

Hamed Taherdoost and Mohsen Saeedi

1.1 INTRODUCTION

In November 2022, the world's population reached eight billion people. Assuming that the population growth rate will continue to decline, according to the medium scenario, the global population is still expected to reach 8.5 billion by 2030, 9.7 billion by 2050, and 10 billion to 12.5 billion by 2100 (Nationen, 2022). This growth has been and will remain the main driver for our unsustainable present and future because the population drives consumption. The amount of consumption per person is also growing.

Several long-term environmental problems are now showing up on a global scale, including pollution of the environment, climate change, loss of biodiversity, deforestation, etc. (Maximillian et al., 2019; Singh & Singh, 2017). The fast growth and dissemination of information and communication technologies have exacerbated the so-called Great Acceleration, which began in the 1950s (Brauch, 2021; Rosa, 2013; Steffen et al., 2015).

In 2012, at the World Conference on Sustainable Development in Rio, the UN member states developed a new set of goals called Sustainable Development Goals (SDGs) to be achieved over 15 years from 2015 to 2030. The SDGs were launched in 2015 after getting inputs, feedback, and views from citizens, civil society organizations, scientists, academics, and the private sector from around the globe. Among 17 SDGs, there are social, economic, and environmental goals. There are a few targets under each goal and some indicators for evaluation.

In 2023, the statistics division of UN DESA (United Nations Department of Economics and Social Affairs) conducted and published a progress assessment. The SDG progress evaluation showed significant challenges. Data analysis from the latest global-level data and custodian agencies revealed concerning results. Among the assessable targets, only 15% were on track to be achieved by 2030. Forty-eight percent of the targets showed moderate or severe deviations from the desired trajectory. Moreover, over 37% of these targets have experienced no progress or regressed below the 2015 baseline values (United Nations Economic and Social Council, 2020).

DOI: 10.1201/9781003609865-1

1.1.1 Sustainable Development and Blockchain

In recent years, "sustainable development" has emerged as a buzzword in the international development sector, with different players applying it in various ways (Horner, 2020; Ogbuoji & Yamey, 2019; Salifu & Salifu, 2024). Beyond environmental sustainability, sustainable development encompasses a multifaceted notion. It also includes aspects of institutions, society, and the economy (Guerrero et al., 2023; Koch et al., 2021). According to Shepherd et al. (2016) and Scopelliti et al. (2018), the concept is likely to remain the dominant development paradigm for a considerable amount of time because it has garnered the kind of widespread attention that other development concepts have not.

Some societal domains highlight the relevance of sustainability in scholarly works. For example, according to the Brundtland Report, sustainability is achieving a state where current and future needs are met through investments, resource exploitation, technological development, and institutional changes (Hariram et al., 2023). This definition highlights the importance of looking at sustainability from a long-term perspective. The necessity for sustainable practices in economic activities is further emphasized by the Cambridge Institute for Sustainability, which emphasizes the creation of "Competitive Sustainability" as a means to a robust recovery and dynamic growth (Javanmardi et al., 2023).

To promote sustainability, individual actions are crucial. People can help lessen their impact on the environment and promote a culture of sustainability by making small changes in their everyday routines. Eating a plant-based diet, reducing water consumption, and switching to more environmentally friendly modes of transportation like biking or public transportation are all examples of what can be done to help the environment (Böhme et al., 2022; Seyfang, 2013).

Societal norms and values also impact sustainable practices. By encouraging a worldview that treats the universe as a machine and places a premium on prediction and control rather than comprehending complicated interactions and relationships, the prevailing social paradigm – also known as the mechanistic paradigm – can impede sustainable lifestyles. In contrast, a relational paradigm can aid in developing sustainable lifestyles by investigating and organizing relational patterns, emphasizing the interdependence of ecological and human systems (Böhme et al., 2022). Sustainable development in all its forms is essential to combating global issues like poverty, inequality, climate change, and environmental degradation, and the SDGs give a blueprint for doing just that (Laininen, 2019).

Innovative technology abounds, giving businesses a leg up in the marketplace (Kamble et al., 2021). The rapid pace of technological development has far-reaching consequences for businesses' environmental consciousness and long-term viability in society (Kouhizadeh & Sarkis, 2018). Blockchain, a distributed ledger technology, is one of the most recent innovations that has a major impact on sustainability (Kouhizadeh & Sarkis, 2018; Taherdoost & Madanchian, 2023).

A distributed ledger system that links data in chronological order within blocks is known as blockchain technology. It has undergone substantial evolution over the years and guarantees data anonymity. Since its introduction by "Satoshi Nakamoto" in a paper about a decentralized Bitcoin system, blockchain technology has expanded to accommodate many more uses. It is currently focusing on applications that need high performance and reliability on a large scale (Dong et al., 2023; Habib et al., 2022; Liu et al., 2023).

One-way blockchain technology works are by eliminating the need for a middleman by enabling numerous nodes to validate transactions in real time. Several industries, including energy, logistics, and education, have begun to use this technology, demonstrating its adaptability and the breadth of its possible influence (Chen et al., 2018; Habib et al., 2022).

Blockchain technology's key features are decentralization, immutability, transparency, and security. This could cause a sea change in many sectors and turn the Internet we know today into an "Internet of Value." There are far-reaching consequences for 21st-century education, corporate operations, and governance due to blockchain's immutable distributed ledger system, which guarantees the authenticity of transactions (Chen et al., 2018; Dong et al., 2023; Kimani et al., 2020).

1.2 BLOCKCHAIN TECHNOLOGY

Blockchain is a distributed ledger that keeps track of transactions across multiple computers in an immutable, decentralized manner. Fifth, computational logic, transparency with pseudonymity, irreversibility of records, peer-to-peer transmission, and distributed databases form its basis (Treiblmaier, 2020).

A distributed ledger system, or blockchain, is a shared database that keeps track of every transaction made by any user and ensures that every user has a copy of the ledger, thanks to replication. A blockchain or hash chain is formed when each block includes a hash value of the header from the block before it, and each block contains a transaction. You can distinguish permissioned blockchains from permissionless ones by looking at how each member's identity is defined in the network (Justinia, 2019).

Blockchain technology's decentralized design ensures that no one entity controls the data and that everyone can independently verify transactions (Namasudra et al., 2021). Everyone with access to the blockchain can see all transactions, thanks to the transparency with pseudonymity principle, and users can choose to be identified by a unique 30-plus character alphanumeric address or by providing proof of identity to others. Every block in a blockchain stores the hash value of the header from the block before it, ensuring that records cannot be changed after a transaction has been entered, according to the principle of irreversibility of records. Blockchain transactions can be effectively programmed using the computational logic principle (Treiblmaier, 2020).

Several industries stand to benefit from blockchain technology's revolutionary potential (Wijesekara & Gunawardena, 2023). These include education, healthcare and biomedical sciences, and knowledge-defined networking. It can be utilized to establish a fair system for evaluating the learning process and its results, guaranteeing their authenticity and offering a reliable method for investing in talent. Concerns about privacy and security and the system's inherent complexity are possible downsides of using blockchain technology in the classroom (Chen et al., 2018).

1.3 KEY FEATURES AND COMPONENTS OF BLOCKCHAIN SYSTEMS

1.3.1 DECENTRALIZATION

Distributed ledger technology underpins blockchain's decentralization efforts (Howell & Potgieter, 2019). This technology also underpins data verification, storage, maintenance, and transmission. The foundation of this system is a network of linked nodes, each of which verifies the integrity of the data stored, thereby maintaining a full record of all transactions that have occurred (Paik et al., 2019; Xu et al., 2019).

Instead of relying on centralized organizations, mathematical methods are used to build trust between distributed nodes. All nodes in the network contribute to the upkeep of consensus algorithms, which allow it to happen. To validate transactions and solve mathematical problems, Bitcoin, for instance, employs a verification mechanism known as PoW. Miners compete with one another. However, the certifier on Ethereum must demonstrate ownership of a specific amount of cryptocurrency using PoS (Hoffman et al., 2020).

Among the many advantages of decentralization are transaction trustworthiness, transparency, and security improvements. It also makes it possible to build DApps and DAOs, autonomous applications, and organizations without a central authority to function (Singh & Kim, 2019). Scalability, energy consumption, interoperability, and regulatory concerns are some challenges that decentralization can bring (Hoffman et al., 2020).

The promise of blockchain technology for decentralization in several domains, such as the Internet of Things (IoT), edge networks, and the IoV, has been the subject of multiple studies (Gadekallu et al., 2021; Prabadevi et al., 2021; Taherdoost, 2023). Chowdhury et al. (2020) offered a broad overview of blockchain technology for Internet decentralization without going into specifics.

1.3.2 IMMUTABLE LEDGER

The immutable ledger is a crucial component of blockchain technology that guarantees data cannot be deleted or altered once recorded. Each transaction is validated and verified before being added to the blockchain, achieving this feature through a consensus mechanism (Yadav et al., 2022). Cryptographic hash functions generate a distinct digital fingerprint for every data block in the blockchain, making the ledger immutable (Rahardja et al., 2021). The hash function connects each block in the blockchain to the one before it, ensuring no changes can be made to the chain without changing the hash value of each block that follows it (Komalavalli et al., 2020; Politou et al., 2019). Because of this, the data recorded on the blockchain cannot be altered, and it is completely secure.

Blockchain technology can revolutionize the financial industry by making transactions more efficient and less fraud-prone (Khadka, 2020; Kimani et al., 2020). Blockchain technology can improve data security and patient privacy in healthcare by building a decentralized and secure system for storing and sharing electronic health records. To improve supply chain efficiency and decrease the likelihood of fraud, supply chain managers can employ blockchain technology to build a transparent and immutable system for monitoring the flow of products and goods (Dutta et al., 2020; Shahnaz et al., 2019). The immutability of blockchain data makes it an ideal system for recording transactions and other events. Industries like healthcare, government, and finance, where the security and integrity of data are paramount, can benefit greatly from this (Dutta et al., 2020).

1.3.3 CONSENSUS MECHANISMS

Consensus mechanisms ensure that every node in a distributed network agrees on the current ledger (Lashkari & Musilek, 2021). They allow for decentralized decision-making and secure the network, making them the backbone of blockchain systems. This text will outline blockchain consensus mechanisms, discussing their uses, classification, and significance.

Crash fault tolerance (CFT) and byzantine fault tolerance (BFT) are two general categories into which consensus mechanisms fall (Yao et al., 2021; Zhou et al., 2023), as classified in Figure 1.1. While BFT mechanisms deal with more complicated malicious activities like purposefully delaying messages and misleading other nodes, CFT mechanisms handle non-malicious faults like delays and losses.

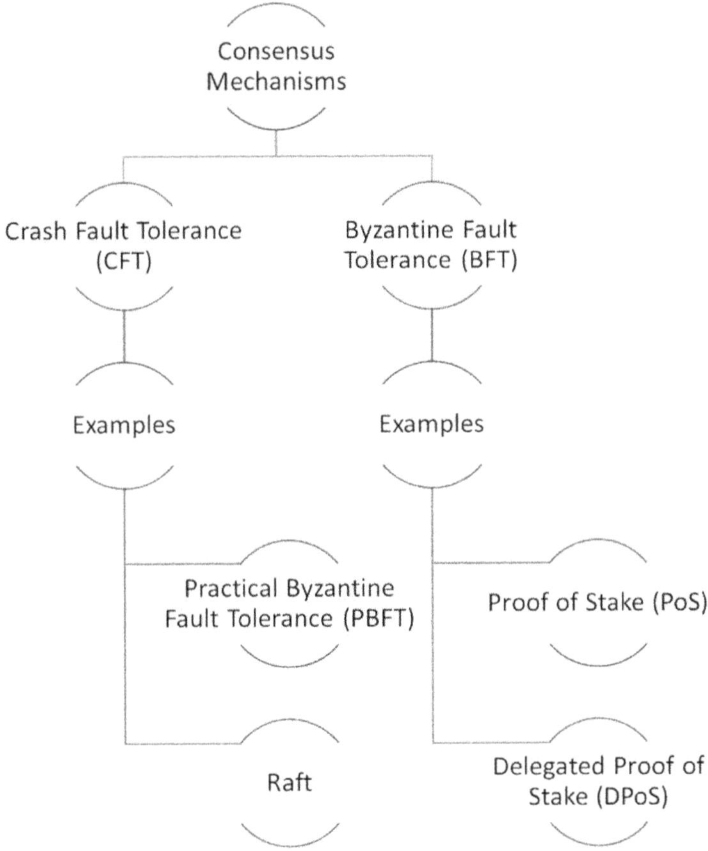

FIGURE 1.1 Consensus mechanisms.

1.3.4 How CFTs Reach Consensus

Non-Byzantine faults are the most prevalent and fundamental distributed system failure, and CFT consensus mechanisms are built to handle them. Some examples of CFT consensus mechanisms include the state machine replication algorithm Practical Byzantine Fault Tolerance (PBFT) (Khan et al., 2022; Zhou et al., 2023), which fixes the original Byzantine Fault-Tolerant algorithm's inefficiency, and the simplicity and reliability-focused consensus algorithm Raft (Zhou et al., 2023).

1.3.5 The Mechanisms for BFT Consensus

Deliberate delays and deceit are examples of more sophisticated malicious actions that BFT consensus mechanisms tackle. Proof of stake (PoS) (Dong et al., 2023; Sayeed & Marco-Gisbert, 2019; Zhou et al., 2023) is one example of a BFT consensus mechanism; it uses stakeholders' stake in the network to select validators; another variation of PoS, delegated proof of stake (DPoS) (Sayeed & Marco-Gisbert, 2019), uses stakeholders' election of delegates to validate transactions; and both use consensus mechanisms.

1.4 SMART CONTRACTS

To facilitate the verification and enforcement of contract negotiations and execution, a digital protocol called a "smart contract" has been developed. Without intermediaries, smart contracts allow for the execution of trustworthy transactions (Wang et al., 2019; Zheng et al., 2020). These deals cannot be undone and can be tracked. A "smart contract is a set of promises specified in digital form, including the protocols within which the parties execute those promises" (Szabo, 1997).

Reduced transaction costs, enhanced security, and increased efficiency are just a few benefits of smart contracts over conventional contracts. Smart contracts automate the execution of contract terms, reducing transaction costs by eliminating intermediaries. Further, a decentralized blockchain network is used to store smart contracts, which makes them more secure and less susceptible to fraud or manipulation. On top of that, smart contracts can cut down on processing and enforcement time and effort, which greatly improves contract execution efficiency (Cheng et al., 2024).

Several problems with smart contracts must be resolved before they can be used, including security, privacy, legal, and performance concerns. Due to a re-entrance vulnerability, the Decentralized Autonomous Organization (DAO) smart contract was the victim of a notoriously malicious attack in 2016 that caused a loss of US$50 million. Another major concern is privacy since smart contracts are published on a blockchain network, which could expose sensitive information. There is a great deal of worry about smart contract legality since this concept still needs to be better defined in many jurisdictions. Smart contracts must be widely adopted, but their performance has significant challenges, such as gas costs and scalability (Khan et al., 2021).

Several exciting developments may soon be coming for smart contracts, which bodes well for their future. Data science smart contracts, which use AI and ML, are believed to greatly enhance smart contracts' functionality and efficiency (Badruddoja et al., 2021). More complicated and dynamic contract structures should be possible once game theory is integrated into smart contracts. Energy management and operations, renewable energy exchange, and peer-to-peer energy trading are some of the possible uses of smart contracts in the energy sector, which is anticipated to experience substantial growth shortly. To enable decentralized and automated processes, several blockchain platforms use smart contracts. A groundbreaking platform, Ethereum, introduced smart contracts, which allowed for the transparent and trustless execution of code in a variety of contexts, including decentralized finance (DeFi) and non-fungible tokens (NFTs) (Vionis & Kotsilieris, 2023).

1.5 CRYPTOGRAPHIC SECURITY

The safety and efficiency of blockchain systems rely heavily on cryptographic methods. Digital encryption, the backbone of blockchain technology, guarantees user data and transaction records' authenticity, integrity, and secrecy. Blockchain systems primarily employ cryptographic methods such as digital signatures, asymmetric cryptosystems, and hash functions. Secure data blocks and a tamper-evident chain are created using hash functions, and secure communication and trust between network participants are enabled by asymmetric cryptosystems (Zhai et al., 2019). Digital signatures are utilized to authenticate transactions and guarantee non-repudiation. The data, network, and consensus layers are just a few of the blockchain infrastructure layers that use these cryptographic techniques to guarantee the system's overall security and integrity (Hardjono & Smith, 2019; Paik et al., 2019; Zhai et al., 2019). Blockchain systems that use these cryptographic techniques have enabled it to build decentralized, secure, and transparent platforms for various uses, such as healthcare data management, supply chain management, and financial transactions (Justinia, 2019).

Using a public key and a private key that corresponds to it, public–private key cryptography is essential for securing transactions. The encryption and decryption procedures made possible by this cryptographic system ensure the privacy and authenticity of data transmitted over the internet, allowing for safe and secure communication. A publicly distributed public key allows anybody to encrypt messages in this architecture, but only individuals with the matching private key can decrypt them and view the original content. By utilizing digital signatures, where the sender signs a message with their private key and the receiver verifies its authenticity with the sender's public key, this asymmetric encryption method secures message content and provides authentication. Data privacy and integrity in digital communications can be assured through public–private key cryptography, which allows for secure transactions (Astorga et al., 2022; Kumar & Tripathi, 2020).

Secure and effective data protection is possible with cryptographic methods like searchable encryption and encryption. The rising use of cloud services has made data storage protection in cloud computing an increasingly pressing issue, necessitating the application of these techniques (Hassan et al., 2022). One of the most basic human rights that cryptography helps to safeguard is the right to privacy, which is jeopardized when cloud providers have unfettered access to user data (Limniotis, 2021). To keep people's and businesses' private information safe, it is essential to use cryptographic security measures to safeguard data and stop unauthorized access.

1.6 TRANSPARENCY AND TRACEABILITY

Thanks to blockchain technology, all transactions are recorded on a public ledger, a decentralized and immutable database that anyone in the network can access, which provides transparency (Sedlmeir et al., 2022). Supply chain management, the fashion and textile industry, and food traceability are just a few of the many sectors interested in blockchain technology for its transparency and traceability features (Azevedo et al., 2023; Badhwar et al., 2023; Ellahi et al., 2023; Sedlmeir et al., 2022). While traceability entails knowing where a product came from, how it was processed, and where it is now after delivery, transparency provides relevant, timely, and trustworthy information in both written and verbal forms (Badhwar et al., 2023; Ellahi et al., 2023; Sedlmeir et al., 2022).

To combat problems like product recalls, fraud, gray markets, stolen goods, and counterfeiting, blockchain technology has become an attractive option for supplying secure traceability and control in supply chains (Ellahi et al., 2023; Jabbar et al., 2021). Supply chain trust and traceability can be improved with blockchain technology because of its immutability, transparency, security, and fault tolerance. Some examples of real-world applications of blockchain technology include the mango traceability pilot project that Walmart ran successfully, the food supply chain transparency initiative that IBM spearheaded with Walmart, the diamond provenance tracking system Everledger, and the integration of blockchain technology for free-range chickens and other foods by Carrefour. In addition to automating the management of outstanding invoices in supply chains, smart contracts powered by blockchain technology can handle sales outstanding amounts that exceed the usual credit contract limit (Jabbar et al., 2021). Nevertheless, scalability and interoperability are two of the most important remaining technical and non-technical obstacles to overcome before blockchain technology is widely used in supply chains (Ellahi et al., 2023; Jabbar et al., 2021; Kouhizadeh et al., 2021).

1.7 INTEGRATING COMPUTATIONAL INTELLIGENCE INTO SUSTAINABLE BLOCKCHAIN SOLUTIONS

An area of artificial intelligence known as computational intelligence centers on creating models and algorithms that mimic biological processes. Examples of such systems include fuzzy logic, neural networks, and evolutionary algorithms (Wagner et al., 2022). Complex problems in many fields, such as engineering, healthcare, and education, can be solved using these methods (Hsieh et al., 2022). Computational intelligence can shape blockchain consensus mechanisms, smart contracts, data privacy, and scalability solutions to promote SDGs (Figure 1.2).

Some examples of how computational intelligence is helping out in the classroom include automating tutoring duties for teachers, identifying instances of conflict in collaborative learning, and providing individualized support to students taking online courses. For example, eTeacher is a system that tracks how students are doing in class and creates a profile for each. Based on that profile, the system can tailor its recommendations for readings and exercises to each individual's needs. By developing smart tutoring systems that can adjust to each student's unique requirements and offer constructive criticism, computational intelligence can elevate the standard of education. These systems can bolster group learning by enabling collaborative writing and employing academically productive speech patterns (Zawacki-Richter et al., 2019).

There are several ways in which computational intelligence, specifically AI and ML, can improve blockchain's sustainability. More efficient and environmentally friendly supply chains are possible with the help of AI and blockchain technology, which can increase supply chain visibility, transparency, and product tracing (Charles et al., 2023). Overcoming scalability issues, AI-ML applications can also aid in managing network traffic and completing a high volume of blockchain transactions. Blockchain technology's immaturity, security, and privacy concerns can be addressed through AI, which improves the technology's architecture and overall performance (Olaniyi et al., 2022).

When applied to the setting of smart cities, AI and blockchain have the potential to enhance social, environmental, and economic sustainability. For example, blockchain technology and Internet of Things (IoT) sensors powered by artificial intelligence can facilitate waste collection, disposal, and recycling. To further a nation's pledge to achieve sustainable development goals, blockchain can supply civil engineers with trustworthy big data to enhance urban resources and services (Rejeb et al., 2021).

FIGURE 1.2 Data science for long-term blockchain sustainability.

1.8 SUMMARY

This chapter examines the convergence of blockchain technology and sustainability within the context of global population growth and escalating environmental challenges. The world population reached eight billion in 2022 and is projected to continue increasing, driving greater consumption and exacerbating sustainability issues such as pollution, climate change, and biodiversity loss. The United Nations' Sustainable Development Goals (SDGs), established in 2015, aim to address these challenges through social, economic, and environmental targets set for 2030. However, recent assessments show significant gaps in achieving these goals, with many targets off track or regressing. The chapter explores the broad concept of sustainable development, which includes environmental, institutional, societal, and economic dimensions, emphasizing the importance of long-term planning and individual actions in fostering sustainability. It also highlights the potential of blockchain technology to enhance transparency, accountability, and efficiency in sustainability efforts. By leveraging blockchain's unique features, the chapter suggests innovative solutions to support the SDGs and address complex global challenges effectively.

REFERENCES

Astorga, J., Barcelo, M., Urbieta, A., & Jacob, E. (2022). Revisiting the feasibility of public key cryptography in light of IIoT communications. *Sensors*, *22*(7), 2561.

Azevedo, P., Gomes, J., & Romão, M. (2023). Supply chain traceability using blockchain. *Operations Management Research*, *16*(3), 1359–1381.

Badhwar, A., Islam, S., & Tan, C. S. L. (2023). Exploring the potential of blockchain technology within the fashion and textile supply chain with a focus on traceability, transparency, and product authenticity: A systematic review. *Frontiers in Blockchain*, *6*, 1044723.

Badruddoja, S., Dantu, R., He, Y., Upadhayay, K., & Thompson, M. (2021). Making smart contracts smarter. *2021 IEEE International Conference on Blockchain and Cryptocurrency (ICBC)*. IEEE.

Böhme, J., Walsh, Z., & Wamsler, C. (2022). Sustainable lifestyles: Towards a relational approach. *Sustainability Science*, *17*(5), 2063–2076.

Brauch, H. G. (2021). Peace ecology in the anthropocene. In *Decolonising conflicts, security, peace, gender, environment and development in the anthropocene* (pp. 51–185). Springer.

Charles, V., Emrouznejad, A., & Gherman, T. (2023). A critical analysis of the integration of blockchain and artificial intelligence for supply chain. *Annals of Operations Research*, *327*(1), 7–47.

Chen, G., Xu, B., Lu, M., & Chen, N.-S. (2018). Exploring blockchain technology and its potential applications for education. *Smart Learning Environments*, *5*(1), 1–10.

Cheng, M., Chong, H.-Y., & Xu, Y. (2024). Blockchain-smart contracts for sustainable project performance: Bibliometric and content analyses. *Environment, Development and Sustainability*, *26*(4), 8159–8182.

Chowdhury, S. H. M., Jahan, F., Sara, S. M., & Nandi, D. (2020). Secured blockchain based decentralised internet: A proposed new internet. *Proceedings of the International Conference on Computing Advancements*. ACM Digital Library.

Dong, S., Abbas, K., Li, M., & Kamruzzaman, J. (2023). Blockchain technology and application: An overview. *PeerJ Computer Science*, *9*, e1705.

Dutta, P., Choi, T.-M., Somani, S., & Butala, R. (2020). Blockchain technology in supply chain operations: Applications, challenges and research opportunities. *Transportation Research Part E: Logistics and Transportation Review, 142*, 102067.

Ellahi, R. M., Wood, L. C., & Bekhit, A. E.-D. A. (2023). Blockchain-based frameworks for food traceability: A systematic review. *Foods, 12*(16), 3026.

Gadekallu, T. R., Pham, Q.-V., Nguyen, D. C., Maddikunta, P. K. R., Deepa, N., Prabadevi, B., . . . Hwang, W.-J. (2021). Blockchain for edge of things: Applications, opportunities, and challenges. *IEEE Internet of Things Journal, 9*(2), 964–988.

Guerrero, O. A., Guariso, D., & Castañeda, G. (2023). Aid effectiveness in sustainable development: A multidimensional approach. *World Development, 168*, 106256.

Habib, G., Sharma, S., Ibrahim, S., Ahmad, I., Qureshi, S., & Ishfaq, M. (2022). Blockchain technology: Benefits, challenges, applications, and integration of blockchain technology with cloud computing. *Future Internet, 14*(11), 341.

Hardjono, T., & Smith, N. (2019). Decentralized trusted computing base for blockchain infrastructure security. *Frontiers in Blockchain, 2*, 24.

Hariram, N., Mekha, K., Suganthan, V., & Sudhakar, K. (2023). Sustainalism: An integrated socio-economic-environmental model to address sustainable development and sustainability. *Sustainability, 15*(13), 10682.

Hassan, J., Shehzad, D., Habib, U., Aftab, M. U., Ahmad, M., Kuleev, R., & Mazzara, M. (2022). The rise of cloud computing: Data protection, privacy, and open research challenges – a systematic literature review (SLR). *Computational Intelligence and Neuroscience, 2022*.

Hoffman, M. R., Ibáñez, L.-D., & Simperl, E. (2020). Toward a formal scholarly understanding of blockchain-mediated decentralization: A systematic review and a framework. *Frontiers in Blockchain, 3*, 35.

Horner, R. (2020). Towards a new paradigm of global development? Beyond the limits of international development. *Progress in Human Geography, 44*(3), 415–436.

Howell, B. E., & Potgieter, P. H. (2019). Governance of blockchain and distributed ledger technology projects: A common-pool resource view. *Workshop on the Ostrom Workshop (WOW6) Conference*, Indiana University Bloomington.

Hsieh, M.-C., Pan, H.-C., Hsieh, S.-W., Hsu, M.-J., & Chou, S.-W. (2022). Teaching the concept of computational thinking: A STEM-based program with tangible robots on project-based learning courses. *Frontiers in Psychology, 12*, 828568.

Jabbar, S., Lloyd, H., Hammoudeh, M., Adebisi, B., & Raza, U. (2021). Blockchain-enabled supply chain: Analysis, challenges, and future directions. *Multimedia Systems, 27*, 787–806.

Javanmardi, E., Liu, S., & Xie, N. (2023). Exploring the challenges to sustainable development from the perspective of grey systems theory. *Systems, 11*(2), 70.

Justinia, T. (2019). Blockchain technologies: Opportunities for solving real-world problems in healthcare and biomedical sciences. *Acta Informatica Medica, 27*(4), 284.

Kamble, S. S., Belhadi, A., Gunasekaran, A., Ganapathy, L., & Verma, S. (2021). A large multi-group decision-making technique for prioritizing the big data-driven circular economy practices in the automobile component manufacturing industry. *Technological Forecasting and Social Change, 165*, 120567.

Khadka, R. (2020). *The impact of blockchain technology in banking: How can blockchain revolutionize the banking industry?* Theseus. https://www.theseus.fi/handle/10024/346030.

Khan, M., den Hartog, F., & Hu, J. (2022). A survey and ontology of blockchain consensus algorithms for resource-constrained IoT systems. *Sensors, 22*(21), 8188.

Khan, S. N., Loukil, F., Ghedira-Guegan, C., Benkhelifa, E., & Bani-Hani, A. (2021). Blockchain smart contracts: Applications, challenges, and future trends. *Peer-to-Peer Networking and Applications, 14*, 2901–2925.

Kimani, D., Adams, K., Attah-Boakye, R., Ullah, S., Frecknall-Hughes, J., & Kim, J. (2020). Blockchain, business and the fourth industrial revolution: Whence, whither, wherefore and how? *Technological Forecasting and Social Change, 161*, 120254.

Koch, D.-J., Vis, J., van der Harst, M., Tendron, E., & de Laat, J. (2021). Assessing international development cooperation: Becoming intentional about unintended effects. *Sustainability, 13*(21), 11571.

Komalavalli, C., Saxena, D., & Laroiya, C. (2020). Overview of blockchain technology concepts. In *Handbook of research on blockchain technology* (pp. 349–371). Elsevier.

Kouhizadeh, M., Saberi, S., & Sarkis, J. (2021). Blockchain technology and the sustainable supply chain: Theoretically exploring adoption barriers. *International Journal of Production Economics, 231*, 107831.

Kouhizadeh, M., & Sarkis, J. (2018). Blockchain practices, potentials, and perspectives in greening supply chains. *Sustainability, 10*(10), 3652.

Kumar, R., & Tripathi, R. (2020). Secure healthcare framework using blockchain and public key cryptography. In K.-K. R. Choo, A. Dehghantanha, & R. M. Parizi (Eds.), *Blockchain cybersecurity, trust and privacy* (pp. 185–202). Springer International Publishing. https://doi.org/10.1007/978-3-030-38181-3_10

Laininen, E. (2019). Transforming our worldview towards a sustainable future. In *Sustainability, human well-being, and the future of education* (pp. 161–200). Palgrave Macmillan.

Lashkari, B., & Musilek, P. (2021). A comprehensive review of blockchain consensus mechanisms. *IEEE Access, 9*, 43620–43652.

Limniotis, K. (2021). Cryptography as the means to protect fundamental human rights. *Cryptography, 5*(4), 34.

Liu, F., He, S., Li, Z., Xiang, P., Qi, J., & Li, Z. (2023). An overview of blockchain efficient interaction technologies. *Frontiers in Blockchain, 6*, 996070.

Maximillian, J., Brusseau, M., Glenn, E., & Matthias, A. D. (2019). Pollution and environmental perturbations in the global system. In *Environmental and pollution science* (pp. 457–476). Elsevier.

Namasudra, S., Deka, G. C., Johri, P., Hosseinpour, M., & Gandomi, A. H. (2021). The revolution of blockchain: State-of-the-art and research challenges. *Archives of Computational Methods in Engineering, 28*, 1497–1515.

Nationen, V. (2022). *World population prospects 2022: Summary of results*. UN.

Ogbuoji, O., & Yamey, G. (2019). Aid effectiveness in the sustainable development goals era: Comment on "'It's about the idea hitting the bull's eye': How aid effectiveness can catalyse the scale-up of health innovations". *International Journal of Health Policy and Management, 8*(3), 184.

Olaniyi, O. M., Alfa, A. A., & Umar, B. U. (2022). Artificial intelligence for demystifying blockchain technology challenges: A survey of recent advances. *Frontiers in Blockchain, 5*, 927006.

Paik, H.-Y., Xu, X., Bandara, H. D., Lee, S. U., & Lo, S. K. (2019). Analysis of data management in blockchain-based systems: From architecture to governance. *IEEE Access, 7*, 186091–186107.

Politou, E., Casino, F., Alepis, E., & Patsakis, C. (2019). Blockchain mutability: Challenges and proposed solutions. *IEEE Transactions on Emerging Topics in Computing, 9*(4), 1972–1986.

Prabadevi, B., Deepa, N., Pham, Q.-V., Nguyen, D. C., Reddy, T., Pathirana, P. N., & Dobre, O. (2021). Toward blockchain for edge-of-things: A new paradigm, opportunities, and future directions. *IEEE Internet of Things Magazine, 4*(2), 102–108.

Rahardja, U., Hidayanto, A. N., Lutfiani, N., Febiani, D. A., & Aini, Q. (2021). Immutability of distributed hash model on blockchain node storage. *Scientific Journal of Informatics*, *8*(1), 137–143.

Rejeb, A., Rejeb, K., Simske, S. J., & Keogh, J. G. (2021). Blockchain technology in the smart city: A bibliometric review. *Quality & Quantity*, 1–32.

Rosa, H. (2013). *Social acceleration: A new theory of modernity*. Columbia University Press.

Salifu, G. A.-N., & Salifu, Z. (2024). Aid administration and sustainable development in post-COVID-19 era in Africa: A review of literature approach. *Cogent Social Sciences*, *10*(1), 2312649.

Sayeed, S., & Marco-Gisbert, H. (2019). Assessing blockchain consensus and security mechanisms against the 51% attack. *Applied Sciences*, *9*(9), 1788.

Scopelliti, M., Molinario, E., Bonaiuto, F., Bonnes, M., Cicero, L., De Dominicis, S., . . . Dedeurwaerdere, T. (2018). What makes you a "hero" for nature? socio-psychological profiling of leaders committed to nature and biodiversity protection across seven EU countries. *Journal of Environmental Planning and Management*, *61*(5–6), 970–993.

Sedlmeir, J., Lautenschlager, J., Fridgen, G., & Urbach, N. (2022). The transparency challenge of blockchain in organizations. *Electronic Markets*, *32*(3), 1779–1794.

Seyfang, G. (2013). Shopping for sustainability: Can sustainable consumption promote ecological citizenship? In *Citizenship, environment, economy* (pp. 137–153). Routledge.

Shahnaz, A., Qamar, U., & Khalid, A. (2019). Using blockchain for electronic health records. *IEEE Access*, *7*, 147782–147795.

Shepherd, E., Milner-Gulland, E., Knight, A. T., Ling, M. A., Darrah, S., van Soesbergen, A., & Burgess, N. D. (2016). Status and trends in global ecosystem services and natural capital: Assessing progress toward aichi biodiversity target 14. *Conservation Letters*, *9*(6), 429–437.

Singh, M., & Kim, S. (2019). Blockchain technology for decentralized autonomous organizations. In *Advances in computers* (Vol. 115, pp. 115–140). Elsevier.

Singh, R. L., & Singh, P. K. (2017). Global environmental problems. In *Principles and applications of environmental biotechnology for a sustainable future* (pp. 13–41). Springer.

Steffen, W., Richardson, K., Rockström, J., Cornell, S. E., Fetzer, I., Bennett, E. M., . . . De Wit, C. A. (2015). Planetary boundaries: Guiding human development on a changing planet. *Science*, *347*(6223), 1259855.

Szabo, N. (1997). Formalizing and securing relationships on public networks. *First Monday*, *2*(9). https://doi.org/10.5210/fm.v2i9.548.

Taherdoost, H. (2023). Blockchain-based internet of medical things. *Applied Sciences*, *13*(3), 1287.

Taherdoost, H., & Madanchian, M. (2023). Blockchain-based e-commerce: A review on applications and challenges. *Electronics*, *12*(8), 1889.

Treiblmaier, H. (2020). Toward more rigorous blockchain research: Recommendations for writing blockchain case studies. In *Blockchain and distributed ledger technology use cases: Applications and lessons learned* (pp. 1–31). Springer.

United Nations Economic and Social Council. (2020). *Sustainable development goals progress chart 2020*. UN: The United Nations. United States of America. Retrieved from https://coilink.org/20.500.12592/q82149 on 3 Dec 2024. COI: 20.500.12592/q82149.

Vionis, P., & Kotsilieris, T. (2023). The potential of blockchain technology and smart contracts in the energy sector: A review. *Applied Sciences*, *14*(1), 253.

Wagner, G., Lukyanenko, R., & Paré, G. (2022). Artificial intelligence and the conduct of literature reviews. *Journal of Information Technology*, *37*(2), 209–226.

Wang, S., Ouyang, L., Yuan, Y., Ni, X., Han, X., & Wang, F.-Y. (2019). Blockchain-enabled smart contracts: Architecture, applications, and future trends. *IEEE Transactions on Systems, Man, and Cybernetics: Systems*, *49*(11), 2266–2277.

Wijesekara, P. A. D. S. N., & Gunawardena, S. (2023). A review of blockchain technology in knowledge-defined networking, its application, benefits, and challenges. *Network, 3*(3), 343–421.

Xu, Y., Ren, J., Zhang, Y., Zhang, C., Shen, B., & Zhang, Y. (2019). Blockchain empowered arbitrable data auditing scheme for network storage as a service. *IEEE Transactions on Services Computing, 13*(2), 289–300.

Yadav, A. S., Agrawal, S., & Kushwaha, D. S. (2022). Distributed ledger technology-based land transaction system with trusted nodes consensus mechanism. *Journal of King Saud University-Computer and Information Sciences, 34*(8), 6414–6424.

Yao, W., Ye, J., Murimi, R., & Wang, G. (2021). A survey on consortium blockchain consensus mechanisms. arXiv preprint arXiv:2102.12058.

Zawacki-Richter, O., Marín, V. I., Bond, M., & Gouverneur, F. (2019). Systematic review of research on artificial intelligence applications in higher education – where are the educators? *International Journal of Educational Technology in Higher Education, 16*(1), 1–27.

Zhai, S., Yang, Y., Li, J., Qiu, C., & Zhao, J. (2019). Research on the application of cryptography on the blockchain. *Journal of Physics: Conference Seriesm, 1168*, 032077.

Zheng, Z., Xie, S., Dai, H.-N., Chen, W., Chen, X., Weng, J., & Imran, M. (2020). An overview on smart contracts: Challenges, advances and platforms. *Future Generation Computer Systems, 105*, 475–491.

Zhou, S., Li, K., Xiao, L., Cai, J., Liang, W., & Castiglione, A. (2023). A systematic review of consensus mechanisms in blockchain. *Mathematics, 11*(10), 2248.

2 Blockchain in Environmental Conservation

*Mohammad Amin Borghei, Reza Ahmadi,
Vahab Esfandani, Sara ravan Ramzani,
and Peter Konhäusner*

2.1 INTRODUCTION

The Fourth Industrial Revolution has changed manufacturing methods and given us alternatives to live higher quality lives resulting in an enormous change in society. Integrating environmental sustainability into the transition is inevitable, though. According to recent studies, Industry 4.0 technology can help corporate sustainability meet its greatest potential and enable excellent sustainable production and green organizations (Dubey et al., 2015; Dubey et al., 2016; Dubey et al., 2017).

Macro-level technological advancement and national sustainability are related (Gouvea et al., 2018); environmental factors in the development of technology improve the welfare of the environment (Song and Wang, 2016). According to Stock and Seliger (2016), the new industrial paradigm facilitates the generation of sustainable industrial value at the micro level by allocating resources in an effective manner through using intelligent value creation modules.

By aligning information technology with organizational objectives, Industry 4.0 tools can improve assessments of environmental sustainability (de Sousa Jabbour et al., 2018). Researchers have studied how Industry 4.0 affects environmental sustainability with particular attention on supply chain management and sustainable manufacturing (Ford and Despeisse, 2016; Jin et al., 2017; Kumar et al., 2018; Luthra and Mangla, 2018). Real-time data synchronization and analysis over-dispersed networks are made possible by blockchain which is a revolutionary method of recording and exchanging data across various ledgers (Natarajan et al, 2017). Governments have used blockchain technology that is already applied (and being applied) in many industries (Centobelli et al., 2021) to improve environmental sustainability (Glavanits, 2020). It makes it easier to monitor pollution and establish green supply chains and manufacture green products (Bai and Sarkis, 2019; Mora et al., 2021; Saberi et al., 2019). Even though blockchain enhances environmental and economic sustainability (Pazaitis et al., 2017; Varsei et al., 2014), research frequently ignores more general environmental issues in favor of concentrating on particular aspects, like green supply chains (Varriale et al., 2020) or city management (Mora et al., 2021).

DOI: 10.1201/9781003609865-2

In order to fully appreciate blockchain technology significance throughout the Fourth Industrial Revolution, it is helpful to comprehend the basic concepts and functions of it initially.

2.2 GENERAL UNDERSTANDING

Blockchain tech employs a network of digital blocks that are accessible to everyone for saving data, functioning as a decentralized recording system to be used in transactions or communication (Moll and Yigitbasioglu, 2019). So as to every single one of these blocks has an electronic signature and a time stamp on it, they are practically immutable (Kokina et al., 2017) (Nakamoto, 2009). The term "blockchain" is rooted in the way that digital blocks are organized into a chain of blocks using a complex mathematical logic known as "hashing" (Nakamoto, 2009; Angelis and Ribeiro da Silva, 2019; Harris and Wonglimpiyarat, 2019).

Blockchain-based applications are being developed at a rapid pace due to the growth of peer-to-peer technology. Bypassing the need for centralized officials to confirm identities, approve transactions, or uphold agreements, this technology allows businesses to share information, services, and products directly. Consequently, this technology accelerates the whole procedure by minimizing the need for intermediaries. First of all, by enabling faster transactions that are distributed and synchronized digitally among several distinct but fewer participants, this may assist businesses and other organizations to increase efficiency and save costs (Davidson et al., 2016). Blockchain technology is most notably used in the creation and management of cryptocurrencies (such as Ethereum and Bitcoin, among others) as well.

Blockchain is being considered in a number of industries other than financial services including supply chain management, taxation, international trade, corporate management, and business activities (Kimani et al., 2020; Pólvora et al., 2020; Centobelli et al., 2021); assessment of blockchain research in management in 2021 revealed how researchers have strategically made the use of blockchain information technologies to achieve competitive edge in a range of industries including supply chain management, artificial intelligence, edge computing, security and privacy, and consortium blockchain. Although the concept of blockchain is still in its inception phases, academics agree that it provides a variety of prospective advantages which will help enterprises fulfill the needs of the Fourth Industrial Revolution (Ahluwalia et al., 2020; Dai and Vasarhelyi, 2017; Moll and Yigitbasioglu, 2019; Hoai Luan et al., 2018; Singh et al., 2019).

Additionally, according to Lee (2019), blockchain will signal a "completely new industrial infrastructure" by drastically changing, if not fully replacing, a number of the present accounting and financial applications. The appeal of blockchain lies in its capacity to facilitate open data exchange, simplifies corporate procedures, reduces operating costs, boosts teamwork, and creates a system that does not require explicit integration of trust in order to function properly, unlike supply chains for example (Francisco and Swanson, 2018). Numerous solutions to challenges with environmental sustainability are offered using the technology of blockchain. This is achieved through three main approaches as stated by Hughes et al. (2019) and Herweijer et al. (2018): preserving resource rights, confirming product origin, and encouraging sustainable practices. Advanced data collecting and analysis techniques might help in the development of new sustainable manufacturing operations. The monitoring and documentation of activities that lead to pollution and environmental deterioration are made possible through these approaches. The prompt development of eco-friendly supply chains and quick decision-making are supported by the real-time collecting and analysis of low-carbon or green data (Bai and Sarkis, 2019; Saberi et al., 2019). A comprehensive review of 30 research papers exploring the intersection of blockchain architecture and supply chain management was carried out by Varriale et al. (2020). Their findings proofed that blockchain has a high potential to significantly enhance environmental sustainability within supply chains. Key benefits identified and include the reduction of CO_2 emissions throughout the supply chain, effective monitoring of hazardous waste exchanges, improved circular economy practices, and the oversight of natural resource utilization particularly within the food and agricultural sector.

On top of that, Glavanits (2020) demonstrates that governments and academics alike have acknowledged blockchain's potential to help achieve the SDGs (Sustainable Development Goals). Glavanits outlines a number of significant efforts in relation to the environmental SDGs. One such scheme is the "Share and Charge" project which was introduced initially in the UK and subsequently extended to the EU. The goal of this effort is to track and manage electric vehicle charging infrastructure using blockchain architecture. He also states how Sacramento, California, is using blockchain system to monitor and control groundwater supplies. The importance of blockchain technology for establishing an environmentally friendly society has recently come to light (Mora et al., 2021). These authors discuss how inordinate numbers of blockchain technological alternatives can promote sustainability from distinct viewpoints depending on the area in which the technology might be employed: administration of resources, city government, and delivery of services. The first two groups of solutions are directly related to challenges of social development given that cities are the most significant ecosystems in which people live and where technology is already playing significant roles. Meanwhile, the third category compiles ideas for resolving interpersonal issues at the level of cities.

Additionally, big data which is commonly used in blockchain and new tech terminology goes to terms like big databases, remote data transfer, artificial intelligence, remote communication, Internet of Things (IoT) devices, various gadgets (such as water meter sensors), data centers, data analysis, and data processing that demands a lot of CPU power. However, blockchain has dispersed networks, certified validators, linked peers, verifiable transactions, consensus mechanisms, decentralization, transparency, and immutability (Hangan et al., 2022).

Previously, a brief introduction and some information have been provided to make it clear in which domain of interest and technology of this study is mainly focused. Having detailed explanations may result in the question of how using blockchain may benefit us in terms of recent contemporary lifestyle. However, it has been cleared indirectly till now. Hereby, some namely advantages are listed and it will then be followed by some of the environmentally focused areas of sustainability for further transparency which could be said to be highly aligned with the topic of interest of this chapter.

2.3 COMMUNITY INVOLVEMENT AND CARBON OFFSET PARTICIPATION

Increased community involvement in environmental conservation initiatives could be facilitated by blockchain. Blockchain-based systems, for instance, can be used to incentivize people and communities to engage in conservation efforts like wildlife monitoring or replanting (Howson, 2019).

To distribute one free SolarCoin to solar energy generators on each MWh of electricity produced, for example, SolarCoin makes use of a blockchain infrastructure. Exchangeable for other digital prizes or as a means of exchange, this prize can be used all over again. Efforts such as Earth Dollar aim to connect blockchain tokens which are representations of specific assets or services on the platform to carbon credits, which are permits for emissions avoided elsewhere.

In order to compute the actual carbon emissions from customer transactions and automatically acquire carbon credits, some particular projects are offering fully computerized payment methods for smart contracts. Veridium Labs, a private firm based in Hong Kong and in collaboration with IBM, is integrating carbon credits generated from Infinite Earth's Forest reserve in Rimba Raya, Central Kalimantan, with their VerdePay payment system (Howson, 2019).

2.4 ENERGY CONSERVATION

Peer-to-peer energy trading networks and other decentralized energy systems can be developed with the use of blockchain technology. By allowing users to exchange excess renewable energy directly with one another, these systems can improve energy efficiency and lessen dependency on fossil fuels (Mengelkamp et al., 2018).

The capacity of blockchain technology to offer traceability and transparency is one of its main benefits. This is especially useful for supply chain management as it tracks everything from the point of origin to the final customer. By ensuring that sustainable practices are strictly followed across the supply chain, this skill supports efforts aimed at promoting environmental sustainability. Blockchain, for instance, can confirm that suppliers of raw materials receive their products responsibly and that businesses follow ecological standards at every stage of manufacturing (Paliwal et al., 2020; Munir et al., 2022).

Blockchain technology can track and manage ecological information securely and with high quality, making it possible to keep transparent and securely records of resource allocation, wildlife trafficking, and conservation initiatives. This capacity improves responsibility and confidence among stakeholders, which is especially useful for monitoring protected areas and conservation funds. For instance, blockchain is being used by firms such as Ocean Eye to enable micropayments for marine conservation operations, offering clear financial incentives to support successful conservation efforts (Insights, 2024; Sanjeev, 2022).

2.5 DISTRIBUTED LEDGER TECHNOLOGY

The definition of blockchain technology according to IBM (2020) is "a shared, immutable ledger for recording transactions, tracking assets, and building trust". When used properly, the four fundamental core elements of blockchain technology (decentralization, immutability, security, and smart contracts) have the potential for offering key advantages without considering the field of industry its being used (Duan et al., 2020).

A collective network of enterprises can monitor assets and document exchanges more easily with the help of the capabilities mentioned. Having access to the distributed file system which offers an irreversible and permanent record of activities called transactions is available to all network participants. The duplication of effort already present in traditional corporate networks can be avoided since this shared ledger only records each transaction once (IBM, 2020).

Furthermore, a supply chain formed based on blockchain offers increased assurance of fair labor standards and human rights. For instance, consumers can be assured that the products they are buying are supplied and made from ethically verified sources when there is a transparent record of the product's history. Smart contracts may be particularly useful for proactively monitoring and managing sustainable terms and regulatory policies as well as for enforcing or managing the necessary corrections and possible changes. Hence, scholars and industry professionals are becoming increasingly interested in sustainable supply chains (Fahimnia et al., 2015).

2.5.1 Immutable Records

Once a transaction in the shared ledger is recorded, no participant may alter or tamper with it. Provided that a transaction record includes an error, this situation calls for addition of another transaction so as to address the issue, after which both transactions can then be displayed.

2.5.2 Smart Contracts

Imagine a world in which there are no intermediaries involved and transactions are carried out seamlessly. Now we will introduce smart contracts: these blockchain-enabled, automated contracts that guarantee quick and effective operations help realize this ambition. Through establishing norms for a range of operations such as corporate bond transfers and travel insurance payments, smart contracts remove the inefficiencies of conventional approaches. Traditional systems are vulnerable to fraud and cyberattacks since they frequently waste money on needless record-keeping as well as external validations. Data verification is further delayed by the transparency challenges as well. The widespread adoption of IoT has resulted in an increase in transactions which exacerbates these issues and slows down corporate procedures. Reduced profitability and an immediate demand for a more reliable solution follow. At this point, blockchain technology enters the picture and has the potential to completely change how we do business.

2.5.3 A Greater Sense of Trust

Within the context of blockchain, a self-executing contract which is specified by specified and precise protocols is called smart contract. As a member of a restricted community, accurate and timely information is assured. Strong security protocols also prevent outside access to your private blockchain data.

2.5.4 ENHANCED SECURITY

All network participants are required to concur on the accuracy of the data. The reason is that approved transactions are permanently stored and cannot be changed or reversed or undone. Besides, there is no chance for any single transaction to be removed by either individuals or even system administrators.

2.5.5 GREATER PERFORMANCE

By adopting an open and shared database which is shared by network members, protracted record reconciliations are avoided. Furthermore, to speed up transactions a smart contract (a set of instructions as already discussed) can be instantly saved on the blockchain and executed (IBM, 2020).

2.6 ROOM FOR A GREATER SUSTAINABILITY

A more comprehensive and broadly applicable perspective on the supply chain has resulted from focusing on the environmental and social aspects of the supply chain in addition to the business aspects, which serve as essentials for sustainable supply chains. The triple-bottom line of sustainability (the economic, social, and environmental bottom lines) may be addressed by blockchain technology potential attributes. As a result, collecting and recognizing instances of sustainable supply chains might serve as an example of the range of applications for blockchain technology.

Blockchain technology can facilitate the collection, storing, and management of data, which result in enabling the support of substantial product and supply chain information. In this scientific context, openness, transparency, neutrality, dependability, and security for all supply chain participants and drivers are possible (Abeyratne and Monfared, 2016).

Ensuring sustainable supply chain standards is an ongoing challenge for the food and beverage segment of business. Merging radio frequency identification (RFID) and blockchain-based technology offers a remarkable approach. The resulting combination makes it easier to create a food supply chain traceability system which enables real-time tracking in compliance with Hazard Analysis and Critical Control Points (HACCP) regulations (Tian, 2016).

It can also record supply chain events in the agricultural sector (Staples et al., 2017).

Since blockchain technology restricts access to authorized parties, it can assist supply chains in identifying dishonest suppliers and counterfeit products, which have the potential to negatively impact society.

A company's supply chain may gain profit economically through implementing blockchain technology in a number of areas that influence its business operations and financial results. In order to make the business case for blockchain technology in the supply chain, we offer a few instances, among many. According to Ward (2007), blockchain technology has the potential to remove intermediaries from the supply chain, resulting in fewer classifications, lower transaction costs, and faster transactions, all of which reduce corporate waste.

Blockchain technology minimizes transaction times and human error while quickly sharing any modifications made to the data. This could lead to the quick implementation of novel products and processes. Blockchain technology can guarantee data security and authenticity, which lowers the cost of protecting against intentional and uncontrolled data change that raises supply chain risks and degrades businesses dependability (Ivanov et al., 2018).

According to Chohan (2019), there is a lot of potential for using blockchain technology in environmental protection and sustainability initiatives. One such use is in the transparency that reduces information asymmetry, which strengthens market mechanisms for resource allocation.

In addition to the above, there are two scenarios in which the development of blockchain technology in smart grids (SGs) impacts environmental issues which recently have been noticed. To begin with, the more the blockchain technology grows, the more renewable energy resources become available. This may reduce the pollution that fossil fuel power plants make. Then, the energy required for blockchain mining increases. Thus, when fossil fuel-based electricity is used to meet this power requirement, carbon emissions will rise as well. As a result, it is necessary to weigh the benefits and drawbacks of applying blockchain technology to the environment (Hakimi et al., 2021).

With a foundational grasp of blockchain's structure and capabilities, we can explore its specific applications, starting with how it revolutionizes supply chain management to promote transparency and sustainability.

To complete the wide scope advantages stated earlier and to dive deeper in to details with the focus of environmental sustainability, there are some particular areas in which blockchain can and also has a positive impact on the environmental sustainability and address it quite fully. So in terms of discussing more, we will have the following diagram and discuss it in more detail.

2.7 MANAGING SUPPLY CHAIN

An inordinate number of people are intended to purchase items that are manufactured morally, yet it can be challenging to find and validate this kind of information. A product crosses through multiple households when it is sold in a store. Companies can simply fabricate information regarding the components and ingredients they use, the locations they discard their waste, how fairly they treat their staff or even the process by which their products are made.

Through increasing supply chain transparency, blockchain technology can be used to monitor products from the producer to the retail location and help stop fraudulent activity, inefficiencies, waste, and unethical behavior. In order to make more ecologically friendly decisions, they can also assist customers in becoming more knowledgeable about the manufacturing and delivery procedures used for each product.

For instance, provided that a type of food was tracked, consumers may buy materials knowing that it was indeed grown locally. Because food would not have to travel so far, carbon emissions would also decrease as a result. Blockchain technology could confirm that a fish sold at a fish market actually came from an ecologically conscious fisherman or that a bag of coffee actually came from a sustainable sector. The Blockchain Development Company is developing an app called Food Trax that uses blockchain technology to track food from farm to store shelf. Another block-chain initiative that seeks to increase supply chain transparency is called Provenace (Thinkers, 2018).

Blockchain technology can help with supply chain environmental issues by enabling traceability and sustainable practice verification. To ensure that the fish are lawfully obtained and free of slave labor the World Wildlife Fund (WWF) has, for example, established a blockchain-based traceability tool for the Pacific tuna business. Customers may confirm the sustainability of their purchases using this system, which tracks tuna from capture to retail using RFID tags and QR codes (Sanjeev, 2022).

Beyond enhancing supply chain transparency, blockchain technology also holds significant promise in transforming recycling initiatives by creating more efficient and incentivized systems.

2.8 RECYCLING

The extant recycling initiatives frequently lack compelling incentives for individuals to take part. Many municipalities bear the burden of managing recycling programs, which leaves many areas devoid of any recycling initiatives at all. Monitoring and measuring the effects of these programs are equally challenging. A blockchain-driven recycling approach can be appealing toward heads for participation through offering a cash payment plan in the form of a cryptographic token for disposing of waste materials such as plastic containers, cans, and bottles. There are currently a number of locations throughout the world with similar arrangements, most notably in Northern Europe. It would be simple to analyze the effects of any location, business, or program participant as well as to openly track metrics like volume, cost, and profit.

A project called Social Plastic, sometimes known as Plastic Bank, collects discarded plastic and exchanges it for cash in terms of services like phone charging or products like cooking fuel at collecting facilities located in developing nations. Their goal is to reduce poverty and clean up the earth from plastic trash. They are now developing an app that will use blockchain technology to enable users to trade physical currency for cryptographic tokens.

Another blockchain-based application called RecycleToCoin is now under development. It will allow users to return used plastic containers via automated equipment in Europe and other countries for a token (Thinkers, 2018).

In addition to optimizing recycling processes, blockchain's potential extends into the energy sector, where it can facilitate decentralized and efficient energy trading networks.

2.9 ENERGY

The centralization of traditional power grids can lead to face inefficiencies in the distribution of energy such as the unused surplus. In regions of the world impacted by poverty or natural catastrophes, grid disruptions may prevent people from having access to energy.

An overline or peer-to-peer blockchain power network would respond to the need to transport electricity over far distances which might result in losses throughout distribution mesh since power can be locally transferred via this kind of trade from a site with excess production to consumption point. Additionally, it would help lessen the demand for storage of power. To solve this problem, Transactive Grid, a joint venture between ConsenSys and LO3 Energy, is creating a blockchain platform.

SunContract is a peer-to-peer energy exchange (trading) network for solar as well as other renewable energy sources powered by blockchain technology. Power plant installations are highly costly and big business corporations or governments frequently foot the bill.

Through directly investing in renewable energy installations locally and globally, businesses and individuals might be able to profit from a blockchain-based platform.

The goal of the blockchain-decentralized application of EcoChain is mainly to establish a marketplace where investors can make green energy investments and take advantage financially out of them. Another blockchain platform called ElectricChain offers a number of applications similar to SolarCoin with the aim of encouraging solar energy installations worldwide (Thinkers, 2018).

While blockchain impact on energy conservation is significant, its ability to ensure compliance and transparency in environmental treaties is equally transformative.

2.10   GREEN AGREEMENTS

The actual implications of environmental agreements might be extremely difficult to monitor and governments and businesses are not always motivated to comply with their commitments. In this regard, fraud and data manipulation are also issues. Since blockchain technology makes it possible to openly monitor crucial ecological data and demonstrate whether commitments were reached, governments and corporations might be prevented from following their ecological promises or disclosing what they have done. When information is released to the generally available blockchain, it is kept permanently. Blockchain-based legal document storage could reduce fraud and manipulation, such as in the worldwide carbon credit scheme. About 979 million is spent a year on system administration alone. There is a small chance that governments and businesses will turn a blind eye to the bribery or illegal sale of carbon credits if you have an unalterable record of the credits that have been purchased and sold (Thinkers, 2018).

Blockchain technology's role in enforcing environmental treaties naturally extends to non-profit organizations, where it can enhance accountability and ensure that contributions are used effectively.

2.11   NON-PROFITS

It might be challenging to follow the money you donate to environmental charities or to see how it is used. In the charity sector bureaucracy, corruption and inefficiency are still ubiquitous. Blockchain technology can make sure that funds meant to be donated to a particular cause or used as a reward for conservation do not end up in the wrong hands due to paperwork and bureaucracy. Money based on blockchain might even be released to the right parties automatically when certain environmental goals are met. This is also another positive point that could be noticed in this domain as well. People in nations without adequate banking infrastructure can benefit from the ability to transfer money without the need for bank accounts, thanks to blockchain technology. This implies that money can be sent to those in need immediately, bypassing a complicated network of intermediaries or a centralized authority. Bitgive and Bithope are two charities working with cryptocurrencies (Thinkers, 2018).

Beyond the realm of non-profits, blockchain can also streamline the implementation and management of carbon taxes, providing a transparent and reliable system for tracking and reducing carbon emissions.

2.12 CARBON TAX

The carbon footprint of each product is not taken into account when determining its environmental impact under the existing approach. This implies that there is little motivation for businesses to market low-carbon products and little reason for customers to buy them. Using blockchain technology to track each product, carbon footprint would protect against manipulation of the data and allow for the calculation of the appropriate carbon tax to be applied at the time of sale. If purchasing a product with a high carbon footprint is more expensive, then consumers will be incentivized to purchase more ecologically friendly products, which will in turn drive corporations to reorganize their supply chains in order to fulfill the market for such items. Each product and firm might receive a score based on the carbon footprint of whatever they sell through a reputation system developed on the blockchain. This would discourage wasteful and environmentally harmful activities and increase transparency in the manufacturing process (Thinkers, 2018).

In the meantime, blockchain architecture makes tracking the products of a certain entity easier. Conversely, conventional systems find it difficult to determine carbon footprint per product. It also helps determine the proper amount of carbon tax for each single and particular business. Supposing a low-carbon product is more expensive and has a bigger carbon impact, customers could nevertheless decide to purchase it. In order to reduce carbon emissions and satisfy customer demand, companies may focus to review, monitor, and restructure their supply chain in response to this new technology and the pressure coming from the market. Blockchain can assist in reducing carbon emissions along the product journey (value chain) by providing the optimal framework for supply chain mapping and adopting low-carbon product design, manufacture, and transportation (de Sousa Jabbour et al., 2018).

2.13 CHANGING INCENTIVES

It can be challenging for individuals as well as businesses to recognize the immediate results of their activities in a world of complexity. As a result, especially in short-term scale, the incentives for acting in an environmentally sustainable way are not always obvious. Blockchain technology can assist people and organizations in observing the true effects of their activities and encourage them to adopt eco-friendly practices. A factory-generated greenhouse gas or waste emissions, a product carbon footprint, or a company's history of complying with ecological framework can all be publicly traced using the blockchain.

The provision of information and the issuance of tokenized credits in exchange for certain acts or reputation systems based on blockchain technology can all serve as incentives for businesses and individuals to act in an environmentally sustainable manner. These new incentives have the potential to drastically change the economic factors which affect us and the future generations who would live in our planet as well (Thinkers, 2018).

These standards may be helpful in validating the sustainability of goods that are promoted as environmentally friendly. Details about the manufacturing of environmentally conscious items can easily become unclear and complicated to determine. People who actually care about the ecosystem might be convinced to buy goods made using a manufacturing process that has been shown to produce negligible greenhouse gas emissions. An example of a sustainable Indonesian forest product is a desk from Ikea. To ensure that the desks are truly made from this particular wood, Ikea must track the wood from the point of cutting through the production process to the finished product. Blockchain technology can help manage this complicated procedure. One such instance is the Endorsement of the Forestry Certification program, which uses blockchain technology to track the origin of almost 740 million acres of certified forests worldwide (Rosencrance, 2017).

In the modern era we are living, in which technology is changing our lives day by day with such an unbelievable paste, thinking about how financially would blockchain affect our lifestyle, and the sustainability of the environment we are surrounded by is remarkably notable. So, next we are analyzing how we are impacted financially with the novel blockchain technology.

Blockchain makes it possible to create cutting-edge financial products or procedures that can assist in environmental preservation simultaneously. Through safe transactions connected to satellite audits, investors may contribute to the protection of, for instance, the forestland through tokenized conservation projects like those created by InvestConservation. This guarantees not only that funds are used wisely to sustain carbon credits and biodiversity (UNEP – UN Environment Programme) in this example but also that conservation initiatives are appropriately monitored and protected (Chohan, 2019).

Notable to remember that this chapter is mainly focused on sustainability in eco-friendly applications of blockchain in general. So, hereafter we are witnessing two different types of analysis based on the regions and industries which have shown more interest in using the blockchain for a better understanding of how this contemporary technology sounds appealing to nations and businesses. Then, we jump into three different business segments and analyze how they perform utilizing blockchain and how they are interconnected closely to the environmental sustainability.

2.14 COUNTRY ANALYSIS

A comprehensive review and examination of interests of different countries have been made in order to clearly understand the global trend in the "blockchain" phenomena. Assigning a paper to a nation or a continent is most definitely challenging.

As a matter of fact, ordinate numbers of papers and articles are written as a result of international cooperation among academics and professionals from several different countries. Nonetheless, it is important to state that when looking at the distribution of articles based on the corresponding author's nationality, China (24%) and the United States (16%) are the two most active nations worldwide.

It is obvious that China is the leading nation. In real life, China has worked on numerous blockchain-related initiatives. For instance, a proposal to launch an exchange-traded fund (ETF) that monitors stocks connected to blockchain technology as underlying assets was recently received by the China Securities Regulatory Commission (CSRC). China Banking News reported that in October 2018, blockchain technology has been applied to regulatory trials in the country's regional equity markets. Due to its perceived ability to address the problems associated with cross-border remittances and monetary inclusion, blockchain is becoming more and more popular in China (Hou et al., 2018). In simpler terms, China seeks to leverage blockchain technology to promote data sharing, increase corporate efficiency, and create better credit systems across a range of industries such as government services, supply chain management, and the Internet of Things (IoT). Blockchain is no longer merely considered a helpful technology for maintaining databases and acquiring new coins, even in the United States. According to Bakarich et al. (2020), the United States has acknowledged the potential expansion of utilizing this technology in the delivery of public services, and numerous initiatives have already reached a high degree of implementation. The first US state to accept Bitcoin for tax payments is Ohio, for instance. However, the states will have to change the current laws if they want the blockchain to firmly establish itself as a "technological imperative" that the public administration cannot ignore.

2.15   INDUSTRY ANALYSIS

Numerous studies concentrate on looking on how the blockchain is being used in specific industries. Energy and utilities make up the most representative sectors (17.3%). The selection of major industrial players is probably responsible for the industries' representativeness. It is interesting to note that the most well-known corporations in the world are currently experimenting with new business models and blockchain-based technology solutions, specifically based on the sectors listed in table. For instance, IBM and Unilever are collaborating in the logistics and supply chain industry to develop a blockchain solution that will streamline Unilever's supply chain and promote transparency, hence establishing a more trustworthy relationship with customers (Venkatesh et al., 2020). In relation to the agricultural and agri-food industry, IBM and Walmart are creating a blockchain program for food safety that will enable tracking of all the processes that the food has gone through using a QR code on the label. The shipper Maersk is creating TradeLens which is a program designed to automate logistical flows in the transportation (airport/maritime) industry by removing bottlenecks that result in significant waste (Suyambu et al., 2020). Knowing that four-fifths of all items used on a daily basis worth a total of $4 trillion are shipped by sea each year is considerable for this report. Trans-shipment costs are predicted to be impacted by 15% to 20%, and on some particular routes like New York to Amsterdam this impact may reach 50%. TradeLens uses blockchain data management to eliminate this waste. Last but not least, Kodak's KodakCoin concept is similarly fascinating. Photographers can save and license their images straight from their blockchain with this blockchain- and cryptocurrency-based business (Corbet et al., 2020).

2.16 AGRICULTURE SEGMENT

As already stated, the blockchain technology holds considerable potential for providing efficient solutions in a number of developmental domains such as transportation, education, health, and agriculture. Mediators have been viewed in these areas as one of the obstacles to work on development. Blockchain technology eliminates the need for middlemen by enabling peer-to-peer business transactions. As a result, client-server-based Internet technologies employing traditional mediator-centered solutions are giving way to blockchain technology. To solve problems unique to a given domain, blockchain technology can be used with other technologies. For instance, the Internet of Things (IoT) and blockchain technology together have the potential to track how limited resources are used including the amount of energy consumed and the level of groundwater. The agriculture sector also is adopting blockchain technology due to a number of business factors. The first and topmost key reason for using blockchain technology in agriculture is to remove deceitful players and low-quality procedures from the food supply chain. For instance, AgriDigital (agridigital.io) is a blockchain-powered cloud-based platform that offers a productive interface for the agricultural supply chain. Establishing customer and retailer trust about the uniqueness and authenticity of food products is the second highly important feature. For instance, ripe.io is a company that offers a blockchain-based platform just for such purposes. For agricultural producers, particularly farmers in economically disadvantaged nations, blockchain technology may open them new markets. To give farmers access to new markets, AgriLedger (agriledger.io) uses distributed ledger technology.

The following is a general classification of the blockchain technology's prominent applications in the agriculture industry.

2.16.1 CROWDSOURCING FOR AGRICULTURAL PRODUCTS DEVELOPMENT

More and more entrepreneurs and start-up businesses are turning to blockchain-based cryptocurrencies or digital tokens to raise capital for their projects. Initial Coin Offering (ICO mechanism) is the overall term for this application mode (Catalini and Gans, 2018). An example of a digital financial instrument is an ICO which can be funded using fiat or cryptocurrency. One such application is N agriCoin (nagricoin.io), which has generated funds to create intelligent fertilizers that promote plant growth.

2.16.2 AGRICULTURAL SUPPLY CHAIN MANAGEMENT DEVELOPMENT

The application of blockchain technology in agricultural supply chain management is motivated by a number of business factors. First, the main driver for the use of blockchain technology for the food supply chain is to ensure that farmers receive their due portion of the food produced. Resolving consumer concerns over food safety ranks as the second most significant factor. As an illustration, the cloud-based supply chain management software AgriDigital mainly supports the grain supply chain. These platforms aim to efficiently link the different players in the supply chain, including farmers, distributors, storage providers, retailers, and customers. The primary features provided by these platforms are the creation of digital assets that represent tangible objects and the recording and tracking of their journey within the supply chain. They put it into practice by offering a collection of smart contracts that run on a blockchain platform like Ethereum.

2.16.3 Monitoring and Accountability

One of the most significant challenges the agricultural and food processing sectors are facing is ensuring the quality of edible products and agricultural commodities (Kamilaris et al., 2019). When it comes to food products, composition, origin, and purity are the terms which consumers demand authenticity for. These expectations are reasonable when foodstuffs are expensive and are somehow consumed far from their place of origin and have a direct impact on the consumers' health and safety. Providing precise and validated details concerning the nutritional content and place of origin of the food product is one way to overcome this issue. This function is precisely fulfilled by blockchain technology when it comes to tracking and tracing the source of food supply. The immutability framework of the blockchain frequently serves as the foundation of blockchain-based solutions to this problem, together with additional technologies like isotope testing, QR coding, RFID tagging, and DNA marking.

2.16.4 Farmer-Focused Financial Solutions Powered by Blockchain

It might be difficult for conventional financial institutions like banks, insurance providers, and cooperative credit entities to incorporate small farmers into the mainstream financial industry. Reflecting this backdrop, farmers are improperly exploited by illicit moneylenders and traders. Blockchain technology which first surfaced as a means of pay with virtual currencies like Bitcoin can be utilized to offer financial services with a focus on agriculture. One such program that COIN22 (coin22.com) created and executed in order to assist small-scale agricultural producers manage the risk of droughts, floods, and low crop yields is the Agri-Wallet (Sylvester, 2019). Additionally, an increasing number of companies as well as government entities are exploring the use of blockchain technology to offer microcredit, farm insurance, and subsidy transfer (Bermeo Almeida et al., 2018).

2.16.5 Underground Water Resources

The primary objective of the circular economy concept is to sustainably link economic activities with environmental welfare (Murray et al., 2017). Investment in digitalization as part of Industry 4.0 provides enormous potential to generate sustainable economic growth while protecting the environment and preserving social responsibility (Beier et al., 2020). Accordingly, water-related data can be safely stored in an immutable database, thanks to blockchain technology. This fundamental aspect of blockchain technology offers a simple framework for stakeholder collaboration in water governance. Smart contracts can help with sustainable water governance as well by codifying economic and environmental factors (Satilmisoglu et al., 2024).

The amount of portable water is limited and gets far lower every day. The earth's useable water is rapidly degrading, which poses a detrimental impact on subsistence. Water conservation is of paramount importance since problems may arise for humans in the future. Tiwari et al. (2020) are the ones who suggested a particular blockchain-based "decentralized water management system" which is a potential approach to address this issue. They propose a blockchain system for the public Ethereum operating system that is intended to preserve water consumption and meets the supply-demand desires of every user in a peer-to-peer network. Water may be used more efficiently along using blockchain technology, allowing every household to lend or borrow surplus or necessary water from any other household members in the network. Having ultimate goal of promoting water conservation, this design may provide the basic for the application of blockchain technology which will be helpful to increase transparency within the water management system to build a system design with the intention of conserving water by fulfilling the "demand and supply" of every user in a peer-to-peer network. Using Ethereum as the platform, a smart contract has been created for transactions (P2P network of ten households).

For customers, a web interface is made. Thus, the foremost objective is to employ blockchain technology to develop an intelligent water management system for ten families, allowing water to be shared among peers based on demand. A peer-to-peer distributed data sorting system is what the blockchain is. As a result of the blocks joining together to form chains, the blockchain is created. A significant number of trades are made by all users who are part of the blockchain's architecture for passing transactions and storing historical data in the form of blocks. It is feasible to play out a rocketed mark on the trade using the puzzle key and the hash work considering each hub has an open key in addition to a cryptographic key. To verify whether the subject of the electronic mark actually signaled the deal, each center point employs the open key (Dogo et al., 2019; Chohan, 2019). This exchange is included in a square with a structure of "chains" that are systematically linked together throughout time after being made at a specific cycle. Exchanges that are not approved cannot be stored in the square since each user has access to their exchange history by verifying the records they own. Three characteristics of blockchain include information integrity, security, and decentralization. From an African viewpoint, Dogo et al. (2019) investigated how blockchain and Internet of Things (IoT) technologies affect intelligent water management systems. This work takes into account the requirements that African countries must typically meet with the implementation of new water management systems. These requirements include continuous monitoring, maintenance prioritization, remote control of water distribution, and transparent and confidential compliance with regulatory and policy requirements.

Before closing the chapter, this is quite remarkable to discuss shortly about the challenges that traditional blockchains may apply on environment. Alike the benefits and applications we called earlier now we are focusing on how potentially could blockchain harm the environment. "Invaluable Challenges". As a result, we are focusing on three different projects (case studies) to further get brighter resolutions over the blockchain sustainable solutions.

2.17 SUSTAINABLE BLOCKCHAIN SOLUTIONS: CASE STUDIES OF POLKADOT, POLYGON, AND CARDANO

2.17.1 Case Study of Polkadot (DOT)

The environmental implications of conventional blockchain systems specifically those that employ proof-of-work (PoW) consensus processes have given rise to considerable concerns. Innovative blockchain initiatives which prioritize efficiency, scalability, and minimal environmental impact have formed up in response, namely Polkadot (DOT).

PoW consensus algorithms that need a lot of processing capacity to solve complicated computational issues are a key aspect of traditional blockchain systems like Bitcoin. The mining process is a major source of carbon emissions due to its massive electricity use. A hunt for more environmentally friendly blockchain technologies has been spurred by the environmental impact of such systems.

2.17.1.1 Polkadot's Environmentally Friendly Approach

2.17.1.1.1 Scalability and Efficiency

The detachment of the consensus process from the state transition mechanism is one of Polkadot primary advances. Compared to traditional blockchain systems that include these services, this architectural separation enables more efficient processing and lowers the overall power resource consumption. Polkadot presents a practically workable approach to achieve core flexibility and scalability without the environmental costs associated with high resource utilization by partitioning these two key components.

2.17.1.1.2 Consensus Mechanism: Proof of Stake

Inspired by Tendermint and HoneyBadgerBFT, Polkadot uses a novel asynchronous Byzantine fault-tolerant (BFT) resolution mechanism. Polkadot consensus approach does not require considerable computing power, in contrast to PoW systems. Instead, a proof-of-stake (PoS) system is employed wherein validators are selected due to their network stake. Using this strategy, Polkadot consensus process is more environmentally friendly since it uses a lot less energy.

2.17.1.1.3 Network Flexibility and Upgradability

Polkadot is intended to be a scalable and completely adaptable blockchain framework. With regard to its design, new technologies can be integrated with ease and no hard forks or excessively complex decentralized coordination are required. By taking this future-proofing measure, Polkadot can make sure that its technology is always strengthened and optimized, which will eventually lead to an even smaller environmental impact. Polkadot's efficient upgrading and adaptability allow it to quickly adopt new green technologies as they become available.

2.17.1.1.4 *Parachains and Shared Security*

Polkadot architecture is unique in that it makes use of parachains those that are independent chains which operate parallel to the main relay chain. There is less need for duplicated processing power with this configuration since it allows for shared security and parallel processing. Polkadot optimizes resource utilization without risking security by integrating security and facilitating insecure interchain transactions. Lower energy use and a lessening of the environmental effect result from this resource-efficient utilization.

2.17.1.1.5 *Governance and Decentralization*

An additional factor in Polkadot environmental sustainability is its governance approach. Given its adaptability, the network can change and grow as a result of decisions made by community members. Changes intended to increase efficiency and lessen environmental effect can be made more easily, thanks to this decentralized approach. To guarantee that the community may vote on and support policies that improve sustainability, the governance system incorporates a referendum mechanism for significant choices (Wood, 2016).

2.17.2 CASE STUDY OF POLYGON (MATIC)

2.17.2.1 Polygon's Environmentally Friendly Approach

2.17.2.1.1 *Scalability and Efficiency*

The architecture of Polygon is built to maximize efficiency and scalability which are two remarkable factors in minimizing environmental impact. Polygon makes it possible to set up an infinite number of chains owning their particular characteristics and configurations through the inclusion of the Staking Layer. Considering the parallel processing capabilities of such configuration, less redundant computational power is required, which lowers energy consumption.

2.17.2.1.2 *Consensus Mechanism: Proof of Stake*

Proof-of-stake (PoS) consensus approach is used by Polygon, and it is naturally more energy efficient than proof-of-work (PoW) systems. Validating parties in PoS is selected according to the quantity of tokens they hold and are ready to issue "stake" as security. Due to the fact that validators are driven to maintain the network integrity without utilizing an excessive amount of electricity, this approach greatly decreases the requirement for processes that consume a lot of energy.

2.17.2.1.3 *Parachains and Shared Security*

Multiple chains or parachains that operate parallel to the main chain are supported by Polygon which is a distinctive feature of its architecture. Since shared security and parallel processing are enabled, resources can be employed efficiently. Polygon minimizes the overall computing load, boosts the network's efficiency, and reduces its environmental impact by pooling security throughout multiple chains.

2.17.2.1.4   Network Flexibility and Upgradability

The network architecture and basis of Polygon are remarkably flexible and able to be upgraded easily. New technologies are also possible to be smoothly incorporated into its infrastructure with no severe forks or complicated decentralized administration. Polygon can continuously improve its system optimization state to achieve superior performance and less environmental impact, and all are possible due to its flexibility and adaptability potentials.

2.17.2.1.5   Governance and Decentralization

Polygon sustainability efforts are facilitated by its executive governance approach. Given that the network is decentralized, community may decide which features require to be optimized and upgraded. This methodology guarantees the seamless implementation of modifications targeted at minimizing environmental effect and increasing efficiency. Consensus establishment and Polygon Funding Proposals (PFPs) are two examples of governance framework strategies that facilitate community driven decision making.

2.17.2.1.6   Community Treasury and Economic Support

Polygon has established a Community Treasury by which it provides continuous support for the network advancement and its expansion. Having a prearranged release of the native POL token, the Treasury guarantees the accessibility of funds for further improvement plans. Through promoting continuous innovation and adaptability or flexibility as a whole to new environmentally friendly technology, this economic system encourages sustainability (Bjelic et al., 2017).

2.17.3   CASE STUDY OF CARDANO (ADA)

Cardano, a company known for its creative thinking and committed sustainability directorship, incorporates environmental concerns into the foundation of its development philosophy and everyday core operations.

2.17.3.1   Proof-of-Stake Consensus

The Ouroboros proof-of-stake (PoS) consensus algorithm developed by Cardano is intended to use less possible energy resources. Ouroboros uses a random selection mechanism to choose validators according to the stake size they own in contrast to energy-intensive mining methods. When comparing this technology to PoW systems, the energy consumption is significantly reduced.

2.17.3.2   Research and Innovation

Cardano places a strong emphasis on peer-reviewed development and thorough academic research to guarantee that all improvements and upgrades are both environmentally friendly and sound academically. By taking the previous approach, the network efficiency and security are made more efficient, and Cardano's long-term viability is accordingly guaranteed.

2.17.3.3 Governance and Community Involvement

Decentralized decision making and solid community involvement are emphasized entirely under the Cardano governance model. Following this inclusive approach, plans and initiatives with a sustainability focus can be integrated, guaranteeing that sustainability issues will play a key role in the network's future.

2.17.3.4 Dedicated Sustainability Efforts

Cardano dedication to environmental responsibility is being witnessed by the selection of a dedicated sustainability director. This position entails managing all sustainability programs and making sure Cardano's growth is in line with international environmental norms and objectives (Hoskinson, 2017).

2.18 SUMMARY

Finally, using blockchain technology into environmental conservation initiatives offers an achievable approach toward strengthening sustainability in multiple domains. Improving the accountability and effectiveness of environmental programs may be achieved in an enormous manner through using blockchain inherent advantages of decentralization, immutability, and transparency. Technology has the potential to be transformative as being witnessed by its capacity to decrease fraudulent activities, boosting processes and ensuring sustainability as well as traceability in supply chains. It can be predicted that as blockchain technology develops further its potential applications will spread in terms of providing innovative solutions to serious issues related to the environment. By highlighting its function in encouraging responsible resource management, improving energy efficiency, and supporting green projects, this chapter has demonstrated how blockchain might contribute to a more sustainable future. In order to achieve global sustainability goals and promote a greener world, blockchain technology is required to be further explored in detail and applied across diverse environmental contexts.

REFERENCES

Abeyratne, S. & Monfared, R. 2016. Blockchain ready manufacturing supply chain using distributed ledger. *International Journal of Research in Engineering and Technology*, 05.

Adams, R., Kewell, B. & Parry, G. 2018. Blockchain for good? Digital ledger technology and sustainable development goals. In *Handbook of sustainability and social science research*, pp. 127–140. Springer, Cham.

Ahluwalia, S., Mahto, R. V. & Guerrero, M. 2020. Blockchain technology and startup financing: A transaction cost economics perspective. *Technological Forecasting and Social Change*, 151, 119854.

Angelis, J. & Ribeiro da Silva, E. 2019. Blockchain adoption: A value driver perspective. *Business Horizons*, 62, 307–314.

Bai, C. & Sarkis, J. 2019. Green supplier development: A review and analysis. In *Handbook on the sustainable supply chain*, pp. 542–556. Elgaronline.

Bakarich, K. M., Castonguay, J. J. & O'Brien, P. E. 2020. The use of blockchains to enhance sustainability reporting and assurance. *Accounting Perspectives*, 19, 389–412.

Beier, G., Ullrich, A., Niehoff, S., Reissi, M. & Habich, M. 2020. Industry 4.0: How it is defined from a sociotechnical perspective and how much sustainability it includes – a literature review. *Journal of Cleaner Production*, 259, 120856.

Bermeo Almeida, O., Cardenas, M., Samaniego-Cobo, T., Ferruzola, E., Cabezas-Cabezas, R. & Bazán-Vera, W. 2018. Blockchain in agriculture: A systematic literature review. *4th International Conference, CITI 2018*, Guayaquil, Ecuador, November 6–9, Proceedings.

Bjelic, M., Nailwal, S., Chaudhary, A. & Deng, W. 2017. *POL: One token for all polygon chains.* Available: https://resources.cryptocompare.com/asset-management/10343/1718808312317. pdf.

Catalini, C. & Gans, J. S. 2018. *Initial coin offerings and the value of crypto tokens.* National Bureau of Economic Research.

Centobelli, P., Cerchione, R., Esposito, E. & Oropallo, E. 2021. Surfing blockchain wave, or drowning? Shaping the future of distributed ledgers and decentralized technologies. *Technological Forecasting and Social Change*, 165, 120463.

Chohan, U. W. 2019. Blockchain and environmental sustainability: Case of IBM's blockchain water management. *The 21st Century (CBRI)*.

Corbet, S., Larkin, C., Lucey, B. & Yarovaya, L. 2020. KodakCoin: A blockchain revolution or exploiting a potential cryptocurrency bubble? *Applied Economics Letters*, 27, 518–524.

Dai, J. & Vasarhelyi, M. A. 2017. Toward blockchain-based accounting and assurance. *Journal of Information Systems*, 31, 5–21.

Davidson, S., Filippi, P. D. & Potts, J. 2016. *Economics of blockchain.* Available: SSRN: https:// ssrn.com/abstract=2744751 or http://dx.doi.org/10.2139/ssrn.2744751.

de Sousa Jabbour, A. B. L., Jabbour, C. J. C., Foropon, C. & Godinho Filho, M. 2018. When titans meet – can Industry 4.0 revolutionise the environmentally-sustainable manufacturing wave? The role of critical success factors. *Technological Forecasting and Social Change*, 132, 18–25.

Dogo, E. M., Salami, A. F., Nwulu, N. I. & Aigbavboa, C. O. 2019. Blockchain and internet of things-based technologies for intelligent water management system. In *Artificial intelligence in IoT*, pp. 129–150. Springer, Cham.

Duan, J., Zhang, C., Gong, Y., Brown, S. & Li, Z. 2020. A content-analysis based literature review in blockchain adoption within food supply chain. *International Journal of Environmental Research and Public Health*, 17, 1784.

Dubey, R., Gunasekaran, A., Papadopoulos, T. & Childe, S. J. 2015. Green supply chain management enablers: Mixed methods research. *Sustainable Production and Consumption*, 4, 72–88.

Dubey, R., Gunasekaran, A., Papadopoulos, T., Childe, S. J., Shibin, K. T. & Wamba, S. F. 2017. Sustainable supply chain management: Framework and further research directions. *Journal of Cleaner Production*, 142, 1119–1130.

Dubey, R. K., Tripathi, V., Dubey, P. K., Singh, H. B. & Abhilash, P. C. 2016. Exploring rhizospheric interactions for agricultural sustainability: The need of integrative research on multi-trophic interactions. *Journal of Cleaner Production*, 115, 362–365.

Fahimnia, B., Tang, C., Davarzani, H. & Sarkis, J. 2015. Quantitative models for managing supply chain risks: A review. *European Journal of Operational Research*, 247.

Ford, S. & Despeisse, M. 2016. Additive manufacturing and sustainability: An exploratory study of the advantages and challenges. *Journal of Cleaner Production*, 137, 1573–1587.

Francisco, K. & Swanson, D. 2018. The supply chain has no clothes: Technology adoption of blockchain for supply chain transparency. *Logistics*, 2, 2.

Glavanits, J. 2020. Sustainable public spending through blockchain. *European Journal of Sustainable Development*, 9, 317.

Gouvea, R., Kapelianis, D. & Kassicieh, S. 2018. Assessing the nexus of sustainability and information & communications technology. *Technological Forecasting and Social Change*, 130, 39–44.

Hakimi, S. M., Hasankhani, A., Shafie-Khah, M., Bisheh Niasar, M. & Asadollahi, H. 2021. Blockchain technology in the future smart grids: A comprehensive review and frameworks. *International Journal of Electrical Power & Energy Systems*, 129.

Hangan, A., Chiru, C.-G., Arsene, D., Czako, Z., Lisman, D. F., Mocanu, M., Pahontu, B., Predescu, A. & Sebestyen, G. 2022. Advanced techniques for monitoring and management of urban water infrastructures – an overview. *Water*, 14, 2174.

Harris, W. L. & Wonglimpiyarat, J. 2019. Blockchain platform and future bank competition. *Foresight*, 21, 625–639.

Herweijer, C., Waughray, D. & Warren, S. 2018. Building block (chain) s for a better planet. *World Economic Forum*. Available: http://www3.weforum.org/docs/WEF_Building-Blockchains.pdf.

Hoai Luan, P., Tran, T. H. & Nakashima, Y. 2018. December. A secure remote healthcare system for hospital using blockchain smart contract. In *2018 IEEE globecom workshops (GC Wkshps)*, pp. 1–6. IEEE.

Hoskinson, C. 2017. Why we are building Cardano. IOHK, Hong Kong, White Paper, 2017. Available: https://whycardano. com.

Hou, J., Wang, H. & Liu, P. 2018. Applying the blockchain technology to promote the development of distributed photovoltaic in China. *International Journal of Energy Research*, 42, 2050–2069.

Howson, P. 2019. Tackling climate change with blockchain. *Nature Climate Change*, 9, 644–645.

Hughes, L., Dwivedi, Y. K., Misra, S. K., Rana, N. P., Raghavan, V. & Akella, V. 2019. Blockchain research, practice and policy: Applications, benefits, limitations, emerging research themes and research agenda. *International Journal of Information Management*, 49, 114–129.

IBM. 2020. *What is blockchain?* [Online]. Available: www.ibm.com/topics/blockchain [Accessed May 15, 2024].

Insights, S. 2024. *Top 10 biodiversity conservation trends in 2024 startus insights*. Available: https://www.startus-insights.com/innovators-guide/biodiversity-conservation-trends/.

Ivanov, D., Dolgui, A. & Sokolov, B. 2018. The impact of digital technology and Industry 4.0 on the ripple effect and supply chain risk analytics. *International Journal of Production Research*, 57, 1–18.

Jin, M., Tang, R., Ji, Y., Liu, F., Gao, L. & Huisingh, D. 2017. Impact of advanced manufacturing on sustainability: An overview of the special volume on advanced manufacturing for sustainability and low fossil carbon emissions. *Journal of Cleaner Production*, 161, 69–74.

Kamilaris, A., Fonts, A. & Prenafeta Boldú, F. X. 2019. The rise of blockchain technology in agriculture and food supply chains. *Trends in Food Science & Technology*, 91, 640–652.

Kimani, D., Adams, K., Attah-Boakye, R., Ullah, S., Frecknall-Hughes, J. & Kim, J. 2020. Blockchain, business and the fourth industrial revolution: Whence, whither, wherefore and how? *Technological Forecasting and Social Change*, 161, 120254.

Kokina, J., Mancha, R. & Pachamanova, D. 2017. Blockchain: Emergent industry adoption and implications for accounting. *Journal of Emerging Technologies in Accounting*, 14, 91–100.

Kumar, G., Subramanian, N. & Maria Arputham, R. 2018. Missing link between sustainability collaborative strategy and supply chain performance: Role of dynamic capability. *International Journal of Production Economics*, 203, 96–109.

Lee, J. Y. 2019. A decentralized token economy: How blockchain and cryptocurrency can revolutionize business. *Business Horizons*, 62, 773–784.

Luthra, S. & Mangla, S. K. 2018. Evaluating challenges to Industry 4.0 initiatives for supply chain sustainability in emerging economies. *Process Safety and Environmental Protection*, 117, 168–179.

Mengelkamp, E., Gärttner, J., Rock, K., Kessler, S., Orsini, L. & Weinhardt, C. 2018. Designing microgrid energy markets: A case study: The Brooklyn microgrid. *Applied Energy*, 210, 870–880.

Moll, J. & Yigitbasioglu, O. 2019. The role of internet-related technologies in shaping the work of accountants: New directions for accounting research. *The British Accounting Review*, 51, 100833.

Mora, H., Mendoza-Tello, J. C., Varela-Guzmán, E. G. & Szymanski, J. 2021. Blockchain technologies to address smart city and society challenges. *Computers in Human Behavior*, 122, 106854.

Munir, M., Habib, S., Hussain, A., Shahbaz, M., Qamar, A., Masood, T., Sultan, M., Abbas, M. M., Imran, S., Hasan, M., Akhtar, M., Ayub, H. M. U. & Salman, C. A. 2022. Blockchain adoption for sustainable supply chain management: Economic, environmental, and social perspectives citation. *Frontiers in Energy Research*, 10, 899632.

Murray, A., Skene, K. & Haynes, K. 2017. The circular economy: An interdisciplinary exploration of the concept and application in a global context. *Journal of Business Ethics*, 140, 369–380.

Nakamoto, S. 2009. *Bitcoin: A peer-to-peer electronic cash system.* Cryptography Mailing list at https://metzdowd.com

Natarajan, Harish, Krause, Solvej & Gradstein, Helen. 2017. *Distributed ledger technology and blockchain.* FinTech Note; No. 1. Washington, DC: World Bank. http://hdl.handle.net/10986/29053 License: CC BY 3.0 IGO.

Paliwal, V., Chandra, S. & Sharma, S. 2020. Blockchain technology for sustainable supply chain management: A systematic literature review and a classification framework. *Sustainability*, 12, 7638.

Parmentola, A., Petrillo, A., Tutore, I. & de Felice, F. 2022. Is blockchain able to enhance environmental sustainability? A systematic review and research agenda from the perspective of Sustainable Development Goals (SDGs). *Business Strategy and the Environment*, 31, 194–217.

Pazaitis, A., de Filippi, P. & Kostakis, V. 2017. Blockchain and value systems in the sharing economy: The illustrative case of backfeed. *Technological Forecasting and Social Change*, 125, 105–115.

Pólvora, A., Nascimento, S., Lourenço, J. S. & Scapolo, F. 2020. Blockchain for industrial transformations: A forward-looking approach with multi-stakeholder engagement for policy advice. *Technological Forecasting and Social Change*, 157, 120091.

Rosencrance, L. 2017. Blockchain technology will help the world go green. *Bitcoin Magazine* [Online]. Available: www.nasdaq.com/articles/blockchain-technology-will-help-the-world-go-green-2017-05-09 [Accessed May 16, 2024].

Saberi, S., Kouhizadeh, M., Sarkis, J. & Shen, L. 2019. Blockchain technology and its relationships to sustainable supply chain management. *International Journal of Production Research*, 57, 2117–2135.

Sanjeev, S. 2022. Using blockchain technology in environmental conservation. *Earth.Org* [Online]. Available: https://earth.org/using-blockchain-technology-in-environmental-conservation/ [Accessed].

Satilmisoglu, T. K., Sermet, Y., Kurt, M. & Demir, I. 2024. Blockchain opportunities for water resources management: A comprehensive review. *Sustainability*, 16, 2403.

Singh, S., Ra, I.-H., Meng, W., Kaur, M. & Cho, G. 2019. SH-BlockCC: A secure and efficient internet of things smart home architecture based on cloud computing and blockchain technology. *International Journal of Distributed Sensor Networks*, 15, 155014771984415.

Song, M. & Wang, S. 2016. Can employment structure promote environment-biased technical progress? *Technological Forecasting and Social Change*, 112, 285–292.

Staples, M., Chen, S., Falamaki, S., Ponomarev, A., Rimba, P., Tran, A. B., Weber, I., Xu, X. & Zhu, J. 2017. *Risks and opportunities for systems using blockchain and smart contracts*. Sydney: Data61 (CSIRO).

Stock, T. & Seliger, G. 2016. Opportunities of sustainable manufacturing in Industry 4.0. *Procedia CIRP*, 40, 536–541.

Suyambu, G. T., Anand, M. & Janakirani, M. 2020. Blockchain – a most disruptive technology on the spotlight of world engineering education paradigm. *Procedia Computer Science*, 172, 152–158.

Sylvester, G. 2019. *E-agriculture in action: Blockchain for agriculture*. Bangkok, Thailand: FAO.

Thinkers, F. 2018. 7 ways blockchain can save the environment and stop climate change. *Future Thinkers* [Online]. Available: https://futurethinkers.org/blockchain-environment-climate-change/ [Accessed May 15, 2024].

Tian, F. 2016. An agri-food supply chain traceability system for China based on RFID & blockchain technology. In *2016 13th International Conference on Service Systems and Service Management (ICSSSM)* (pp. 1–6). IEEE.

Tiwari, S., Gautam, J., Gupta, V. & Malsa, N. 2020. Smart contract for decentralized water management system using blockchain technology. *International Journal of Innovative Technology and Exploring Engineering*, 9, 2046–2050.

Varriale, V., Cammarano, A., Michelino, F. & Caputo, M. 2020. The unknown potential of blockchain for sustainable supply chains. *Sustainability*, 12, 9400.

Varsei, M., Soosay, C., Fahimnia, B. & Sarkis, J. 2014. Framing sustainability performance of supply chains with multidimensional indicators. *Supply Chain Management: An International Journal*, 19, 242–257.

Venkatesh, V. G., Kang, K., Wang, B., Zhong, R. Y. & Zhang, A. 2020. System architecture for blockchain based transparency of supply chain social sustainability. *Robotics and Computer-Integrated Manufacturing*, 63, 101896.

Ward, T. 2007. *Blockchain could help us save the environment: Here's how* [Online]. Available: https://futurism.com/blockchain-could-help-save-environment-heres-how [Accessed May 22, 2024].

Wood, G. 2016. Polkadot: Vision for a heterogeneous multi-chain framework. *White Paper*, 21(2327), 4662.

3 Smart Contracts and Sustainable Business Models

Hamed Taherdoost

3.1 INTRODUCTION

By decreasing transaction costs, boosting transparency and trust, and encouraging social-proof behaviors, businesses can implement sustainable business models more successfully, thanks to blockchain technology's ability to enable smart contracts (Dal Mas et al., 2020; Mercuri et al., 2021). A case study that looks at the blockchain-based platform Devoleum demonstrates how smart contracts can be used to make long-term business plans (Mercuri et al., 2021).

A project's ability to remain sustainable over its entire life from an economic, environmental, and social standpoint may be aided by smart contracts. Smart contracts can be integrated with other technologies, including the Internet of Things (IoT), during the operating phase to provide safe data storage, improve efficiency and safety, and monitor environmental issues like pollution and energy use (Cheng et al., 2024). Implementing uniform criteria and well-defined procedures is crucial for the sustainability of smart contracts (Groschopf et al., 2021).

Further research is required to fully understand the potential of smart contracts in developing sustainable business models. It is recommended that case studies and empirical research be carried out to understand better how smart contracts contribute to creating social value. Viable models for this capacity must be validated to demonstrate that smart contracts can produce cumulative capacities for controlling social, environmental, and economic performance (Cheng et al., 2024). It will take an interdisciplinary investigation to fully understand smart contracts' environmental and economic impacts (Groschopf et al., 2021).

3.2 SUSTAINABILITY IN BUSINESS MODELS

The idea of a business model provides an abstract representation of the value flow and the interactions between the several value components that comprise an organizational unit. Important components of organizations focused on the critical value elements are the proposal, development, delivery, and capturing of value. The success of any business depends critically on having an easy way to communicate the relationship and function of different parts (Chesbrough, 2010).

DOI: 10.1201/9781003609865-3

A business model illustrates how a business operates and makes money. This is so because any organization can only survive in the long run with money, especially earnings. Therefore, business models are crucial in defining a company's success and show how well management handled the obstacles presented by the (broad) competitive business environment. Though the logic of this framework has not changed, the problems that organizations face today – along with their increasing complexity – and the paradigm shift that has taken place in the assessment of organizational success – in light of these problems – are what is truly revolutionary in the modern world (Ogrean & Herciu, 2020).

Higher profits are thought to result from business model breakthroughs than from product or process innovations (Lindgardt et al., 2012). Higher resilience and risk reduction may also be advantages of sustainable business models (Choi & Wang, 2009). As more companies realize these advantages, they show a growing interest in implementing sustainable ideas (Awan & Sroufe, 2022; Evans et al., 2009). Revenue models, one of the most important parts of business models, must be constructed to enable expansion and sustainability. Aspects of the business plan, such as the area under cultivation, can significantly affect income (Remeňová et al., 2020).

3.3 IMPORTANCE

Organizations that adopt sustainable business practices will experience substantial financial gains. Bocken et al. (2014) identified numerous sustainable business model archetypes. The previous patterns, which encompass value recovery from waste, material, and energy efficiency optimization and functionality provision as opposed to ownership, have the potential to generate cost savings and generate new sources of revenue. Customers, investors, and society ought to be the central concerns of sustainable business models, according to Evans et al. (2017). This can result in competitive advantages and long-term profitability.

Schaltegger et al. (2016) found that organizations that integrate sustainability into their fundamental business strategy are more inclined to achieve enhanced financial performance and gain a competitive edge over those that perceive sustainability as an afterthought. This assists businesses in differentiating their products and services, reducing expenses, and attracting environmentally aware investors and customers. In their study, Gimpel et al. (2020) proved that businesses can achieve financial benefits by incorporating sustainability principles into their operational frameworks, especially when rival firms embrace sustainable practices to compete.

Including environmental concerns in the value proposition, value creation and delivery systems, and value capture systems of the firm is one of the most crucial components of sustainable business models (Ferlito & Faraci, 2022; Nosratabadi et al., 2019). This requires a rethink of goods and services to reduce their negative environmental impacts and to provide benefits to stakeholders.

To this end, one strategy is to include concepts from a green economy and a circular economy into the business model (Najmaei & Sadeghinejad, 2023). This includes designing products meant to last a long time, be reused, remanufactured, and recycled, and convert to renewable and recyclable materials. By putting into practice creative business models like product-service systems, which permit access rather than ownership, one can reduce resource consumption and waste.

Developing a sustainable business model requires a holistic, systemic approach considering the interdependencies between the company, its stakeholders, and the larger ecosystem (Ferlito & Faraci, 2022). Furthermore, strategic choices must be made to integrate sustainability into the core values and approach of the company (Knudson, 2023). Organizations that connect their business plans with environmental sustainability can make long-term profits and lessen their environmental impact simultaneously.

One way to include social responsibility in a conventional business plan is to launch corporate social responsibility (CSR) projects. CSR means going above and beyond legal obligations to address societal and environmental issues (Del Baldo & Baldarelli, 2017). For CSR to create shared value and guarantee long-term sustainability, it must be included in the fundamental company strategy and value proposition (Del Baldo & Baldarelli, 2017; Hu et al., 2020).

Green and sustainable business models (GnSBMs) research has expanded significantly during the last 20 years, moving from a multidisciplinary to an interdisciplinary subject. GnSBMs have historical roots in groundbreaking research on stakeholder theory in the 1980s, environmental science in the 1960s, and business strategy in the 1950s. These works were published in the ensuing decades. Analyzing the life cycle shows that research on GnSBMs underwent a first phase between 2002 and 2013, following which it began to increase exponentially and is expected to do so until about 2040 (Najmaei & Sadeghinejad, 2023).

Sustainable business model innovation is a change in a company's activities intended to either increase benefits or lessen negative environmental and societal effects (Minatogawa et al., 2022). Sustainability ideas must be included in the company's basic operating framework according to the sustainable business model innovation paradigm. Value should be prioritized in any sustainable business model innovation framework, using a multi-stakeholder process and thoroughly examining each aspect of the business plan (Ferlito & Faraci, 2022; Velter et al., 2022). Businesses must increasingly innovate their business models by putting changes, improvements, and replacements in various organizational components to be profitable in a dynamic environment (Ferlito & Faraci, 2022).

Sustainability has become a major competitive advantage for companies within the last few years. Businesses can set themselves apart from their competitors and succeed over time by including environmentally friendly business practices (Abbasi Kamardi et al., 2022; Jiao et al., 2023; Tarnovskaya, 2023). Using green intellectual capital might help one to attribute sustainability to a competitive advantage. Integrating environmental consciousness into a company's intangible assets, such as knowledge and skills, may increase operational efficiency, and more environmentally friendly products that meet customer demands may be offered. Moreover, reducing environmental effects and lowering expenses might give the business a competitive advantage (Jiao et al., 2023).

The application of sustainable supply chain techniques is one more crucial element. Companies that align their commercial operations with supply chain sustainability objectives can develop competitiveness and establish a sustainable competitive advantage. Because of this proactive attitude to managing changes in the market and environmental issues, firms can operate successfully in dynamic business environments (Sun et al., 2022). Figure 3.1 summarizes the many benefits of sustainable business concepts.

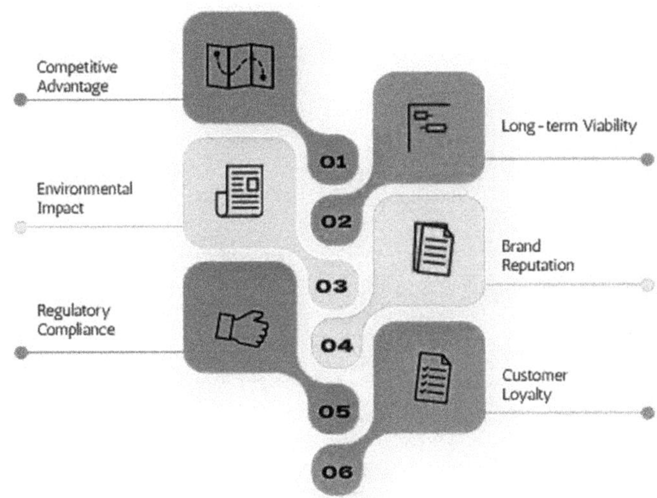

FIGURE 3.1 The value of sustainability for businesses.

3.4 DEFINITION AND COMPONENTS OF SUSTAINABLE BUSINESS MODELS

The essence of sustainable business models is "a simplified representation of the elements, the interrelation between these elements, and the interactions with its stakeholders that an organizational unit uses to create, deliver, capture, and exchange sustainable value for, and in collaboration with, a broad range of stakeholders" (Geissdoerfer et al., 2018).

Unlike traditional business models, sustainability-based models prioritize sustainability concepts as the main motivating element above other considerations. They also combine social, environmental, and commercial activities to provide value for society and customers. To accomplish their sustainability goals, sustainability-based models take a long-term approach and actively manage and innovate with many stakeholders.

The idea of sustainability-based models and the strong sustainability perspective are closely related; the former values ecological features more highly than economic benefits, and the latter views generated and natural capital as mutually incompatible. This contradicts the weak sustainability perspective, which allows some substitution between produced and natural capital.

As stated by Upward and Jones (Upward & Jones, 2016), really sustainable business models are those that simultaneously "tri-profit" creation or generate "positive environmental, social, and economic value throughout their value network, thereby sustaining the possibility that human and other life can flourish on this planet forever". Strong sustainability may be unachievable for companies, and a more realistic approach that recognizes the need for some substitution between created and natural capital is frequently necessary. Sustainable business models create customer and social value by integrating social, environmental, and business activities (Figure 3.2). They have a global market perspective, considering the development of new industrialized countries and the need for more sustainable products and services.

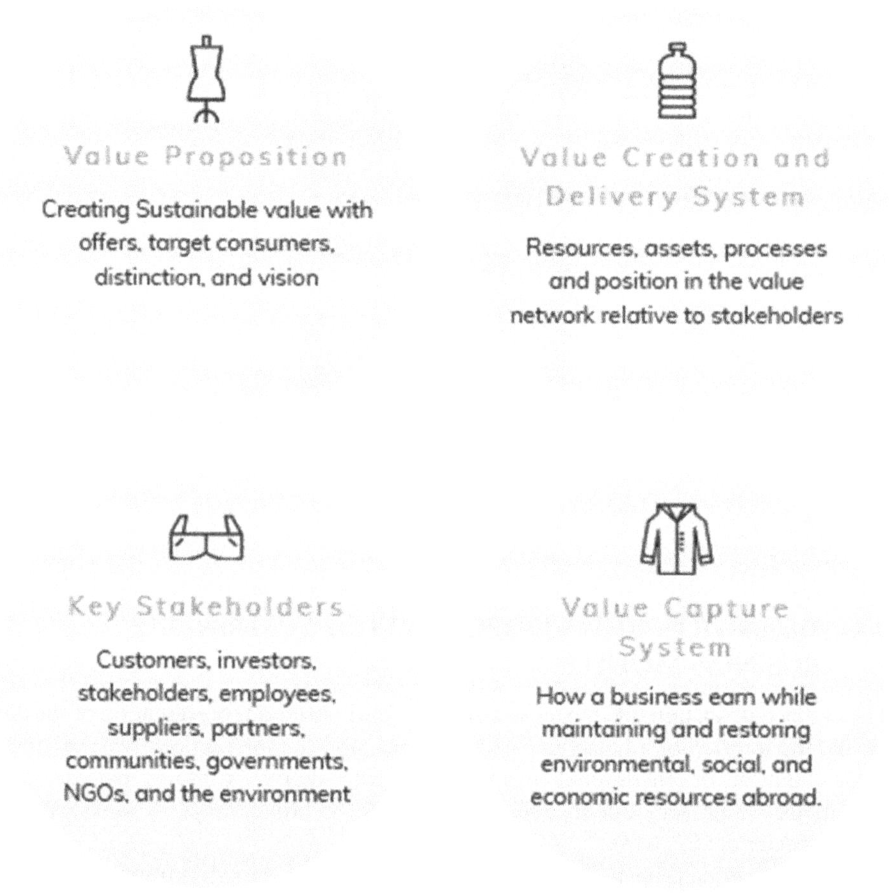

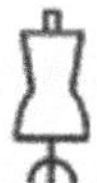

Value Proposition
Creating Sustainable value with offers, target consumers, distinction, and vision
Value Creation and Delivery System
Resources, assets, processes and position in the value network relative to stakeholders
Key Stakeholders
Customers, investors, stakeholders, employees, suppliers, partners, communities, governments, NGOs, and the environment
Value Capture System
How a business earn while maintaining and restoring environmental, social, and economic resources abroad.

FIGURE 3.2 Key components of sustainable business models.

3.5 CHALLENGES IN IMPLEMENTING SUSTAINABLE PRACTICES IN BUSINESS

Businesses need more resources and knowledge to embrace sustainable practices. Small- and medium-sized enterprises are especially affected. Small- and medium-sized enterprises may need more time, money, and experience to design and implement sustainability reporting and procedures. Due to hefty costs and long-term commitment, their operations may need to improve. Large organizations have more sustainability resources and experience than small- and medium-sized enterprises. Balance between stakeholder demands is another key issue. Every stakeholder group has sustainability reporting informational criteria that enterprises may need help to achieve. Businesses often may only discuss their performance and sustainability

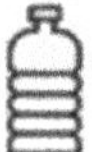

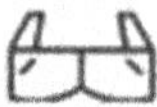

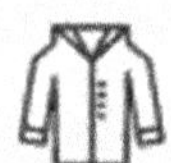

plans through sustainability reports. If there are business concerns, this communication may fail (Setyaningsih et al., 2024).

A further issue many companies encounter is the pervasive short-term focus in many industries. Pressure to make rapid money can make one forget about the long-term and intertemporal trade-offs required for real sustainability. Companies that are more focused on generating short-term profits than on creating long-term value are less likely to make strategic investments in community partnerships, employee training, and research and development necessary for long-term viability. The biggest challenge facing companies wishing to adopt sustainable practices is getting over their short-term thinking (Bansal & DesJardine, 2014).

The need for standards and precise rules for reporting on sustainability presents problems for businesses. Key performance and sustainability indicators come in a wide range, making it challenging for small- and medium-sized enterprises to stay up with the reporting standards. Stability in the sustainability reporting process requires the institutionalization and regular application of managerial commitment, employee involvement, organizational learning, and distribution. A well-defined sustainability strategy and business model are needed to make it easier for companies to implement sustainable practices (Kassem & Trenz, 2020; Lüdeke-Freund, 2009; Setyaningsih et al., 2024).

3.6 SMART CONTRACTS AND SUSTAINABILITY

Smart contracts are self-executing software that allows two parties to automate the terms of an agreement. They are maintained and copied on a blockchain network to provide immutability, security, and transparency (Taherdoost, 2023). Smart contracts do away with the need for intermediaries, lowering the possibility of fraud or mistakes during contractual dealings.

The main traits of smart contracts are dissemination, determinism, and automation. Smart contracts can leverage automation to perform predefined tasks without requiring human interaction when particular requirements are met. Determinism assures that smart contracts will always generate the same output with the same properties given the same input. Distributed smart contracts replicate themselves among all the nodes comprising a blockchain network. Therefore, they are hard to break or have downtime (Bottoni et al., 2020).

Smart contract applications include supply chain management, stock and real estate trading, stock market management, loan providing, dispute resolution, and healthcare. By offering a secure and transparent environment in which to carry out commercial agreements, they enable parties who might not know one another to have reliable contacts (Bottoni et al., 2020). Nevertheless, smart contracts have several disadvantages, including the inability to be changed once implemented, the reliance on the caliber of the code created by the programmer, and the potential for exploitable flaws. The environmental, social, and economic effects of smart contracts and regular contracts are compared in Table 3.1.

TABLE 3.1

The effects of smart contracts versus conventional contracts

Aspect	Environmental Impact	Social Impact	Economic Impact
Smart Contracts	Reduce paper usage and waste	Enhance transparency and trust	Decrease transaction costs
Traditional Contracts	Contribute to deforestation and waste	Limited accessibility and inclusivity	Higher administrative and legal expenses

3.7 HOW SMART CONTRACTS CAN ENHANCE SUSTAINABILITY IN BUSINESS MODELS

Smart contracts can be used in automated compliance, a possible approach to guarantee adherence to laws and regulations related to sustainability. As defined by Szabo (1997), smart contracts are digital agreements that may execute themselves and autonomously enforce the terms of a contract when specific predefined conditions are met. Within the sustainability framework, smart contracts can be built to track and confirm adherence to specific standards, such as waste management, water use, or greenhouse gas emissions, and to carry out predetermined actions or impose penalties in case violations happen (Huckle et al., 2016).

Using renewable energy sources is one sustainability criterion that smart contracts might include. Smart contracts can be designed (Andoni et al., 2019) to track how much energy a company or project uses. They ensure that a predefined percentage of the energy used comes from renewable sources, such as wind or solar power. The smart contract can automatically impose financial penalties or limit resource access until the issue is fixed if the company falls short of the set renewable energy objective.

Throughout the supply chain, smart contracts can trace the origin and path of products or raw materials to ensure they adhere to sustainability criteria like fair trade or organic certification (Tian, 2016). To guarantee that only compliant products reach the market, the smart contract can stop a product from being sold or initiate a recall if it does not satisfy the necessary standards.

Blockchain technology allows smart contracts, which may increase supply chain openness. The self-executing digital agreements facilitate monitoring the flow of resources and products across the supply chain. They also present an auditable and reliable operational record (Saberi et al., 2019). Supply chain stakeholders can obtain real-time information about goods' origin, transport, and status using smart contracts. This raises general efficiency and lowers the possibility of fraudulent conduct (Kshetri, 2018).

Smart contracts allow products and resources to be traceable from origin to the end user, creating a tamper-evident digital trail (Tian, 2016). Every transaction or ownership change is documented on the blockchain, producing an unchangeable record that parties permitted to view it can access (Esmaeilian et al., 2020). Such a degree of transparency reduces the possibility of counterfeiting, helps to ensure that products are genuine, and enables the identification of probable supply chain bottlenecks or issues (Queiroz et al., 2020).

Smart contracts might significantly decrease the need for intermediaries in several transactions, lowering prices and lessening the transactions' environmental impact. Smart contracts make it less necessary to have outside parties monitor and authenticate transactions. Such scenarios are unnecessary since smart contracts automate the implementation of contractual agreements based on predetermined conditions.

Smart contracts can enable direct peer-to-peer energy trading industry transactions between producers and consumers. Doing business makes it possible to do away with conventional intermediaries like energy brokers and exchanges. This follows increased market openness, lower transaction costs, and general energy supply efficiency. Similarly, contracts might automate the verification procedure for the fair trade certification industry. This would reduce the need for outside certifying organizations and guarantee that manufacturers receive just compensation for their products.

The decrease matches the environmental benefits of smart contracts in terms of the number of intermediaries that can be accomplished. Smart contracts may lower carbon emissions associated with office and transportation operations by expediting transactions and reducing the need for paperwork and in-person meetings. Furthermore, smart contracts' increased efficiency and transparency may create better resource distribution and less waste across various businesses.

Smart contracts can automatically gather, handle, and report data from sensors and devices. An open and tamper-proof environmental performance record is the outcome (Christidis & Devetsikiotis, 2016). Real-time monitoring makes it possible to identify issues early on and implement remedies, eventually raising operations' general sustainability (Huckle et al., 2016).

Making well-informed decisions and promoting ongoing progress in sustainability projects require prompt access to trustworthy data (Kamilaris et al., 2019). Real-time monitoring allows companies to track their environmental effect, identify improvement areas, and assess how well their internal sustainability initiatives work. When decision-makers can access real-time updated data, they can maximize resource allocation, react swiftly to changing situations, and uphold environmental requirements (Saberi et al., 2019). In addition, the ability to identify patterns and trends made possible by this timely data allows businesses to take preemptive action to deal with any issues before they worsen (Kshetri, 2017).

Real-time reporting of environmental measures and performance indicators is required to promote ongoing progress in sustainability programs (Kouhizadeh et al., 2019). Companies may foster engagement, responsibility, and openness in their sustainability initiatives by routinely informing stakeholders of their achievements and the challenging issues they encounter (Esmaeilian et al., 2020). By benchmarking against industry standards and best practices, real-time reporting also helps companies identify their areas of strength and need for development (Saberi et al., 2018). This continuous feedback loop urges companies to set high goals, develop innovative solutions, and refine their sustainability plans (Esmaeilian et al., 2020).

One interesting approach to encourage stakeholders to act sustainably is to include incentive mechanisms in smart contracts. Smart contracts are digital contracts that can be made with rewards and limitations to promote desired actions. They are stored on a blockchain and operate automatically (Cong & He, 2019). Smart contracts can help solve the collective action issue and encourage people and companies to support social responsibility and environmental protection. To do this, financial incentives are matched with sustainability objectives.

One kind of incentive system is tokenized incentives. One sustainable behavior that can be rewarded with digital tokens is recycling, cutting back on energy use, or helping with neighborhood cleanups (Nowiński & Kozma, 2017). Digital tokens can be issued as a reward for finishing sustainable activities through smart contracts. These tokens may be exchanged for fiat money to purchase environmentally friendly items or enjoy unique benefits. Using tokenized awards gives stakeholders concrete motivation to engage in sustainability initiatives and can monitor their development over time.

However, smart contracts can also contain penalties for unsustainable behavior. For example, one smart contract that controls a supply chain might automatically take tokens away from providers who violate environmental regulations or engage in unethical behavior (Kouhizadeh et al., 2019). The smart contract would be the one to put this process into practice. The possibility of financial penalties incentivizes suppliers to give sustainability and openness top priority in their operations. Smart contracts can also be customized to reward officials reporting violations, further enhancing the incentive mechanism.

It takes careful consideration of the specific situation and the people concerned to establish incentive schemes through smart contracts successfully. The kind of sustainability goals, the size and interconnection of the system, and the pre-existing societal norms and regulations are only a few of the many aspects that Nowiński & Kozma (2017) stress are essential considerations. Using pilot projects and iterative refining techniques, incentive structures can be optimized and ensured to be efficient in encouraging sustainable behavior. The more the blockchain technology develops, the more likely the smart contracts will be to expedite sustainability by introducing new incentive systems.

3.8 CHALLENGES AND LIMITATIONS

3.8.1 Legal and Regulatory Concerns

Smart contracts raise important legal and regulatory issues that need to be resolved, even if they can greatly improve the management of sustainable supply chains. One major issue that can counter laws like the European Union's General Data Protection Regulation (GDPR) is the immutable character of smart contracts made possible by blockchain technology. This rule requires destroying personal data and granting people the "right to be forgotten". The decentralized trust paradigm may be undermined by governments considering the prospect of controlling blockchain technology (Khan et al., 2021).

Many smart contracts depend on outside "off-chain" databases and oracles to provide information. Among the difficulties this raises are the creation of possible failure points and the emergence of questions about the accuracy of the data. The legality and enforceability of smart contracts still need to be discovered since different jurisdictions have different laws and regulations governing them (Khan et al., 2021; Vionis & Kotsilieris, 2023). Research is still going strong on the problem of quantifying and automating the modeling of legal concepts and contract provisions.

Notwithstanding these difficulties, smart contracts offer the potential to support more sustainable business models by allowing transparent and auditable transactions and by automatically adhering to sustainability standards (Groschopf et al., 2021; Salmerón-Manzano & Manzano-Agugliaro, 2019). Blockchain technology and smart contracts can help circular economy concepts by tracking resources and enabling automatic payments (Groschopf et al., 2021; Kumar & Chopra, 2022). Applications of smart contracts being studied in the energy industry include demand response, electric car charging, and peer-to-peer energy trading.

COVID-19 is forcing a digital revolution in the legal sector, but success will depend on having strong, creative business models that leverage new technology like smart contracts (Hongdao et al., 2022). Legal and regulatory frameworks will need to undergo major adjustments to provide legal clarity and protect consumer interests while promoting innovation. Building governance models and design patterns for smart contracts that balance the interests of all parties requires research that crosses several fields (Khan et al., 2021; Wang et al., 2019).

3.8.2 TECHNOLOGICAL BARRIERS

While smart contracts can facilitate the development of sustainable business models, their effective application will necessitate the resolution of specific technological obstacles (Table 3.2). Scalability is a significant barrier that needs to be overcome because smart contracts rely on blockchain networks. The current blockchain systems' incapacity to manage high throughput and big transaction volumes complicates supply chain processes. If smart contracts are to power viable business models broadly, sharding, off-chain transactions and layer-2 solutions will need to be used to boost scalability.

One of the major hurdles is writing safe and error-free code for smart contracts. However, smart contracts cannot be altered once implemented, so mistakes or flaws could have fatal results. The need for more advanced programming languages, development tools, and formal verification methods to guarantee the security and accuracy of smart contracts has been brought to light by smart contract hacks and exploitation, which have garnered significant media attention. Lowering the risks associated with smart contracts would also need standardizing auditing protocols and best practices.

Integrating smart contracts with the present business processes and data sources requires more work. To operate independently, smart contracts require access to real-world information and events; however, maintaining a safe connection to off-chain data sources can be challenging. The oracles that make this connection possible present a single point of failure and a centralization risk. Data integrity standards and decentralized Oracle networks are necessary for smart contracts to function reliably with the outside world. These technological barriers must be removed to fully realize the potential of smart contracts and enable viable business models.

TABLE 3.2

Challenges in using technology for smart contracts in long-term business plans

Technological Barrier	Impact on Sustainability	Financial Implications
Interoperability	Hindering collaboration	Increased development costs
Scalability	Limiting scalability of sustainable solutions	Increased transaction fees
Security	Risking financial and reputational damage	Loss of investor trust
Complexity	Increasing development time and costs	Higher insurance premiums
Oracle Integration	Limiting applicability of smart contracts	Increased litigation risks
Upgradability and Maintenance	Hindering long-term sustainability	Loss of business opportunities

3.9 ADOPTION HURDLES

Smart contracts and sustainable business models need to get past many challenges. One of the main challenges that have to be solved is the need for standards and compatibility among the many blockchain platforms and smart contract languages. This makes creating and implementing smart contracts that can operate smoothly across several platforms more difficult for companies. Moreover, companies thinking of using the technology may be discouraged by the intricacy of the smart contract code and the potential for unanticipated consequences brought on by flaws or vulnerabilities.

Another barrier that has to be surmounted is the need for legal and regulatory clarity about the application of smart contracts. The legal enforceability of smart contracts is still a highly unsolved issue, especially regarding international transactions. Businesses might employ smart contracts with caution until they are more certain of how the law will see them. Furthermore, a few nations need to adapt their laws faster to reflect new technologies like blockchain and smart contracts.

Especially for small- and medium-sized enterprises, the high initial costs and degree of technical expertise required to create smart contracts and viable business models might deter adoption. Businesses could spend much money on staff training, infrastructure upgrades, and new procedure creation to leverage these technologies. Should they need convincing proof of a significant return on investment, some businesses would be reluctant to undertake these investments.

3.10 FUTURE OUTLOOK

Considering the global economy's continuous effort to address sustainability issues, smart contracts offer a realistic route for deeper integration into sustainable business models. These blockchain-encoded self-executing contracts may revolutionize several business operations by boosting accountability, efficiency, and transparency. Smart contracts automate and simplify the processes of procuring raw materials, production, and distribution, helping businesses guarantee that their supply chains follow sustainable practices. For example, smart contracts can be set up to verify the authenticity and provenance of raw materials, guaranteeing that these goods are obtained ethically and ecologically.

Applying the ideas of the circular economy could be made simpler with the use of smart contracts. Reuse, recycling, or repurposing of resources is done in this paradigm to lessen waste generation and its detrimental effects on the ecosystem. By using smart contracts, businesses can monitor the lifespan of materials and goods, which helps them build closed-loop systems that reduce waste and promote resource efficiency and more environmentally friendly products and packaging. Using smart contracts, which facilitate peer-to-peer trading of renewable energy credits and certificates, consumers can be empowered to directly support renewable energy projects and be encouraged to adopt clean energy sources. Smart contracts also make it easier to include renewable energy sources into already-existing power networks, optimizing energy output and consumption while reducing reliance on fossil fuels.

Beyond the control of renewable energy markets and supply chains, smart contracts encourage environmentally friendly investing and finance methods. Smart contracts match capital allocation with environmental, social, and governance (ESG) standards to encourage investment in sustainable projects and activities. Automated execution of financial instruments associated with sustainability, such as impact investments and green bonds, enables this.

3.11 EMERGING TRENDS AND TECHNOLOGIES SHAPING THE FUTURE OF SUSTAINABILITY

More exact resource allocation and forecasting are becoming feasible because of artificial intelligence (AI) developments. Less waste and more productivity follow from this in supply chains and operations. Artificial intelligence-powered systems may examine vast amounts of data to find trends and opportunities for optimization, helping companies lessen their environmental effect while simultaneously growing their influence.

The Internet of Things (IoT) is a major component of the initiatives to promote sustainability since it allows for real-time monitoring and control of environmental factors. Networked sensors and equipment make businesses' real-time energy, water, and pollution monitoring possible. This helps businesses to take proactive steps to lessen environmental issues and to make more informed decisions. In addition, as companies look for more environmentally friendly substitutes for conventional linear production processes, accepting circular economy ideas – which aim to reduce waste and optimize resource efficiency – is picking up speed. The goals of these concepts are waste reduction and resource efficiency optimization. Companies can redesign their products and operations to prioritize recycling, remanufacturing, and reuse, reducing their dependence on few resources and their environmental effect. When all is considered, the meeting point of these technologies and trends has much potential to boost efforts to advance sustainability and bring about positive change in many sectors.

3.12 OPPORTUNITIES FOR INNOVATION AND COLLABORATION

There are many more chances for creativity and teamwork in the field of sustainable business models that are reinforced by smart contracts. Building new apps and platforms that leverage smart contracts to solve specific sustainability-related issues is one of the biggest potential areas. Companies may collaborate to create decentralized markets where carbon offsets or renewable energy certificates could be traded; this would be made feasible by developing smart contract technology. Through traceability and transparency, these systems enable companies to assess their operations' environmental effects rapidly.

Environmentally friendly supply networks may be innovative because of industry collaboration. Companies who use smart contracts may guarantee that they are using fair labor standards, track the sources of their products, and speed up their procurement processes. Working together, businesses in different industries can develop interoperable smart contract solutions that address shared sustainability issues across industries. Further fostering innovation is collaboration with academic institutions and research organizations, combining industry expertise with state-of-the-art studies in supply chain management, sustainability, and blockchain technology.

3.13 SUMMARY

In recent times, smart contracts have emerged as a potent instrument with the potential to augment the enduring sustainability of business models. Smart contracts enhance accountability, transparency, and efficiency by automating and simplifying processes. Furthermore, they possess the capability to reduce costs and improve flexibility. Intelligent contracts are being implemented in supply chains, energy trading, and carbon offset markets, among other places, to facilitate environmentally favorable processes.

Nevertheless, constraints and challenges must be surmounted, encompassing apprehensions regarding legal and regulatory matters, technological limitations, and criteria for adoption. If organizations want to completely leverage smart contracts for sustainability, they must proactively address these obstacles and involve other relevant parties. Smart contracts may become increasingly ingrained in environmentally sustainable business models in the coming years. Smart contracts assist businesses in maintaining their flexibility and ingenuity, essential for addressing urgent social and environmental issues. This is the case because technological advancements and new opportunities are perpetually changing.

REFERENCES

Abbasi Kamardi, A., Amoozad Mahdiraji, H., Masoumi, S., & Jafari-Sadeghi, V. (2022). Developing sustainable competitive advantages from the lens of resource-based view: Evidence from IT sector of an emerging economy. *Journal of Strategic Marketing*, 1–23.

Andoni, M., Robu, V., Flynn, D., Abram, S., Geach, D., Jenkins, D., . . . Peacock, A. (2019). Blockchain technology in the energy sector: A systematic review of challenges and opportunities. *Renewable and Sustainable Energy Reviews, 100*, 143–174.

Awan, U., & Sroufe, R. (2022). Sustainability in the circular economy: Insights and dynamics of designing circular business models. *Applied Sciences, 12*(3), 1521.

Bansal, P., & DesJardine, M. R. (2014). Business sustainability: It is about time. *Strategic Organization, 12*(1), 70–78.

Bocken, N. M., Short, S. W., Rana, P., & Evans, S. (2014). A literature and practice review to develop sustainable business model archetypes. *Journal of Cleaner Production, 65*, 42–56.

Bottoni, P., Gessa, N., Massa, G., Pareschi, R., Selim, H., & Arcuri, E. (2020). Intelligent smart contracts for innovative supply chain management. *Frontiers in Blockchain, 3*, 535787.

Cheng, M., Chong, H.-Y., & Xu, Y. (2024). Blockchain-smart contracts for sustainable project performance: Bibliometric and content analyses. *Environment, Development and Sustainability, 26*(4), 8159–8182.

Chesbrough, H. (2010). Business model innovation: Opportunities and barriers. *Long Range Planning, 43*(2–3), 354–363.

Choi, J., & Wang, H. (2009). Stakeholder relations and the persistence of corporate financial performance. *Strategic Management Journal, 30*(8), 895–907.

Christidis, K., & Devetsikiotis, M. (2016). Blockchains and smart contracts for the internet of things. *IEEE Access, 4*, 2292–2303.

Cong, L. W., & He, Z. (2019). Blockchain disruption and smart contracts. *The Review of Financial Studies, 32*(5), 1754–1797.

Dal Mas, F., Dicuonzo, G., Massaro, M., & Dell'Atti, V. (2020). Smart contracts to enable sustainable business models: A case study. *Management Decision, 58*(8), 1601–1619.

Del Baldo, M., & Baldarelli, M.-G. (2017). Renewing and improving the business model toward sustainability in theory and practice. *International Journal of Corporate Social Responsibility, 2*, 1–13.

Esmaeilian, B., Sarkis, J., Lewis, K., & Behdad, S. (2020). Blockchain for the future of sustainable supply chain management in Industry 4.0. *Resources, Conservation and Recycling, 163*, 105064.

Evans, S., Gregory, M., Ryan, C., Bergendahl, M. N., & Tan, A. (2009). *Towards a sustainable industrial system: With recommendations for education, research, industry and policy.* University of Cambridge, Institute for Manufacturing.

Evans, S., Vladimirova, D., Holgado, M., Van Fossen, K., Yang, M., Silva, E. A., & Barlow, C. Y. (2017). Business model innovation for sustainability: Towards a unified perspective for creation of sustainable business models. *Business Strategy and the Environment, 26*(5), 597–608.

Ferlito, R., & Faraci, R. (2022). Business model innovation for sustainability: A new framework. *Innovation & Management Review, 19*(3), 222–236.

Geissdoerfer, M., Vladimirova, D., & Evans, S. (2018). Sustainable business model innovation: A review. *Journal of Cleaner Production, 198*, 401–416.

Gimpel, H., Graf-Drasch, V., Kammerer, A., Keller, M., & Zheng, X. (2020). When does it pay off to integrate sustainability in the business model? – a game-theoretic analysis. *Electronic Markets, 30*(4), 699–716.

Groschopf, W., Dobrovnik, M., & Herneth, C. (2021). Smart contracts for sustainable supply chain management: Conceptual frameworks for supply chain maturity evaluation and smart contract sustainability assessment. *Frontiers in Blockchain, 4*, 506436.

Hongdao, Q., Bibi, S., Mu, D., Khan, A., & Raza, A. (2022). Legal business model digitalization: The post COVID-19 legal industry. *SAGE Open, 12*(2), 21582440221093983.

Hu, B., Zhang, T., & Yan, S. (2020). How corporate social responsibility influences business model innovation: The mediating role of organizational legitimacy. *Sustainability*, *12*(7), 2667.

Huckle, S., Bhattacharya, R., White, M., & Beloff, N. (2016). Internet of things, blockchain and shared economy applications. *Procedia Computer Science*, *98*, 461–466.

Jiao, X., Zhang, P., He, L., & Li, Z. (2023). Business sustainability for competitive advantage: Identifying the role of green intellectual capital, environmental management accounting and energy efficiency. *Economic Research-Ekonomska istraživanja*, *36*(2).

Kamilaris, A., Fonts, A., & Prenafeta-Boldú, F. X. (2019). The rise of blockchain technology in agriculture and food supply chains. *Trends in Food Science & Technology*, *91*, 640–652.

Kassem, E., & Trenz, O. (2020). Automated sustainability assessment system for small and medium enterprises reporting. *Sustainability*, *12*(14), 5687.

Khan, S. N., Loukil, F., Ghedira-Guegan, C., Benkhelifa, E., & Bani-Hani, A. (2021). Blockchain smart contracts: Applications, challenges, and future trends. *Peer-to-Peer Networking and Applications*, *14*, 2901–2925.

Knudson, H. (2023). Business models for sustainability. In *Business transitions: A path to sustainability: The CapSEM model* (pp. 101–112). Springer International Publishing.

Kouhizadeh, M., Sarkis, J., & Zhu, Q. (2019). At the nexus of blockchain technology, the circular economy, and product deletion. *Applied Sciences*, *9*(8), 1712.

Kshetri, N. (2017). Can blockchain strengthen the internet of things? *IT Professional*, *19*(4), 68–72.

Kshetri, N. (2018). 1 blockchain's roles in meeting key supply chain management objectives. *International Journal of Information Management*, *39*, 80–89.

Kumar, N. M., & Chopra, S. S. (2022). Leveraging blockchain and smart contract technologies to overcome circular economy implementation challenges. *Sustainability*, *14*(15), 9492.

Lindgardt, Z., Reeves, M., Stalk Jr., G., & Deimler, M. (2012). Business model innovation: When the game gets tough, change the game. In *Own the future: 50 ways to win from the Boston Consulting Group* (pp. 291–298). Wiley Online Library.

Lüdeke-Freund, F. (2009). *Business model concepts in corporate sustainability contexts: From rhetoric to a generic template for 'business models for sustainability'*. Centre for Sustainability Management (CSM), Leuphana Universität Lüneburg.

Mercuri, F., della Corte, G., & Ricci, F. (2021). Blockchain technology and sustainable business models: A case study of Devoleum. *Sustainability*, *13*(10), 5619.

Minatogawa, V., Franco, M., Rampasso, I. S., Holgado, M., Garrido, D., Pinto, H., & Quadros, R. (2022). Towards systematic sustainable business model innovation: What can we learn from business model innovation. *Sustainability*, *14*(5), 2939.

Najmaei, A., & Sadeghinejad, Z. (2023). Green and sustainable business models: Historical roots, growth trajectory, conceptual architecture and an agenda for future research – a bibliometric review of green and sustainable business models. *Scientometrics*, *128*(2), 957–999.

Nosratabadi, S., Mosavi, A., Shamshirband, S., Zavadskas, E. K., Rakotonirainy, A., & Chau, K. W. (2019). Sustainable business models: A review. *Sustainability*, *11*(6), 1663.

Ogrean, C., & Herciu, M. (2020). Business models addressing sustainability challenges – towards a new research agenda. *Sustainability*, *12*(9), 3534.

Queiroz, M. M., Telles, R., & Bonilla, S. H. (2020). Blockchain and supply chain management integration: A systematic review of the literature. *Supply Chain Management: An International Journal*, *25*(2), 241–254.

Remeňová, K., Kintler, J., & Jankelová, N. (2020). The general concept of the revenue model for sustainability growth. *Sustainability*, *12*(16), 6635.

Saberi, S., Kouhizadeh, M., & Sarkis, J. (2018). Blockchain technology: A panacea or pariah for resources conservation and recycling? *Resources, Conservation and Recycling, 130*(March), 80–81.

Saberi, S., Kouhizadeh, M., Sarkis, J., & Shen, L. (2019). Blockchain technology and its relationships to sustainable supply chain management. *International Journal of Production Research, 57*(7), 2117–2135.

Salmerón-Manzano, E., & Manzano-Agugliaro, F. (2019). The role of smart contracts in sustainability: Worldwide research trends. *Sustainability, 11*(11), 3049.

Schaltegger, S., Hansen, E. G., & Lüdeke-Freund, F. (2016). Business models for sustainability: Origins, present research, and future avenues. *Organization & Environment, 29*(1), 3–10.

Setyaningsih, S., Widjojo, R., & Kelle, P. (2024). Challenges and opportunities in sustainability reporting: A focus on small and medium enterprises (SMEs). *Cogent Business & Management, 11*(1), 2298215.

Sun, J., Sarfraz, M., Khawaja, K. F., & Abdullah, M. I. (2022). Sustainable supply chain strategy and sustainable competitive advantage: A mediated and moderated model. *Frontiers in Public Health, 10*, 895482.

Szabo, N. (1997). Formalizing and securing relationships on public networks. *First Monday, 2*(9). https://doi.org/10.5210/fm.v2i9.548

Taherdoost, H. (2023). Smart contracts in blockchain technology: A critical review. *Information, 14*(2), 117.

Tarnovskaya, V. (2023). Sustainability as the source of competitive advantage: How sustainable is it? In *Creating a sustainable competitive position: Ethical challenges for international firms* (pp. 75–89). Emerald Publishing Limited.

Tian, F. (2016). An agri-food supply chain traceability system for China based on RFID & blockchain technology. *2016 13th International Conference on Service Systems and Service Management (ICSSSM).*

Upward, A., & Jones, P. (2016). An ontology for strongly sustainable business models: Defining an enterprise framework compatible with natural and social science. *Organization & Environment, 29*(1), 97–123.

Velter, M., Bitzer, V., & Bocken, N. (2022). A boundary tool for multi-stakeholder sustainable business model innovation. *Circular Economy and Sustainability, 2*(2), 401–431.

Vionis, P., & Kotsilieris, T. (2023). The potential of blockchain technology and smart contracts in the energy sector: A review. *Applied Sciences, 14*(1), 253.

Wang, S., Ouyang, L., Yuan, Y., Ni, X., Han, X., & Wang, F.-Y. (2019). Blockchain-enabled smart contracts: Architecture, applications, and future trends. *IEEE Transactions on Systems, Man, and Cybernetics: Systems, 49*(11), 2266–2277.

4 Blockchain Integration in Renewable Energy

Hamed Taherdoost and Mitra Madanchian

4.1 INTRODUCTION

Blockchain technology has emerged as a transformative force across various industries, and its potential to revolutionize the sustainable energy sector is gaining significant attention (Wang & Su, 2020). This chapter explores the role of blockchain technology in advancing a sustainable energy future, specifically focusing on integrating renewable energy sources such as solar and wind power into the grid.

Blockchain technology is a decentralized and transparent platform that enables secure and efficient transactions without intermediaries. It operates on a distributed ledger system, where all participants can access the same transaction data, eliminating the need for confirmations, actualization of volumes, and many forms of reconciliation. This transparency and security save time, reduce costs, and minimize human error, making blockchain an attractive solution for various industries, including the energy sector (Andoni et al., 2019).

In the context of climate change and the depletion of non-renewable resources, the importance of renewable energy cannot be overstated. Renewable energy sources such as solar, wind, and hydroelectric power offer a sustainable alternative to fossil fuels, reducing greenhouse gas emissions and promoting energy independence. However, integrating renewable energy into the existing energy infrastructure poses challenges due to the intermittent nature of these sources (Anees, 2012). Blockchain technology can help address these challenges by optimizing energy distribution, promoting renewable energy generation, and actively empowering consumers to participate in the energy market (Baashar et al., 2021).

Blockchain-based smart contracts can automate energy transactions, enabling real-time monitoring, control, and optimization of energy flows within the grid. This has the potential to improve grid management, enable demand response programs, and facilitate the integration of renewable energy sources into the existing energy infrastructure (Pfeifer et al., 2018). Moreover, blockchain enables decentralized energy systems, allowing individuals and organizations to generate, consume, and trade energy directly with each other. This peer-to-peer energy trading can optimize energy distribution, promote renewable energy generation, and empower consumers to participate in the energy market (Soto et al., 2021).

Integrating blockchain technology in the renewable energy sector offers significant potential for advancing a sustainable energy future. By enabling the creation of decentralized energy markets, blockchain can facilitate integrating renewable

DOI: 10.1201/9781003609865-4

energy sources into the grid, promote energy efficiency, and empower consumers. As the global energy landscape continues to transform, blockchain technology will play an increasingly important role in shaping the future of sustainable energy.

4.2 FUNDAMENTALS OF BLOCKCHAIN TECHNOLOGY

Blockchain technology is a decentralized, distributed ledger system that enables secure and transparent transactions without intermediaries. Its architecture consists of three main components: the network, consensus, and application layers. The network layer is responsible for communication between nodes on the network. It includes protocols such as TCP/IP, HTTP, and WebSockets, which allow nodes to send and receive data across the network (Asim, 2017). The consensus layer ensures that all nodes on the network agree on the state of the blockchain and includes consensus algorithms such as proof of work (PoW), proof of stake (PoS), and delegated proof of stake (DPoS) (Yang et al., 2019).

The consensus algorithms are used to prevent malicious actors from manipulating the data and ensure that all nodes on the network agree on the state of the blockchain. The application layer is where the actual blockchain applications are developed and deployed, including smart contracts, decentralized applications (DApps), and other blockchain-based services. Cryptography is used to secure the data on the blockchain by encrypting it and providing a digital signature to verify the authenticity of the data (Wu et al., 2021).

The decentralized nature of the network makes it more resilient and resistant to attacks or failures, and the distributed nature of the network prevents a single point of failure, making it difficult for hackers to take down the network. Blockchain technology has the potential to revolutionize industries by providing a secure, transparent, and efficient way to transfer value and data (Javaid et al., 2021). It can improve operational efficiency and save costs significantly, making it an attractive solution for various applications, including supply chain management, healthcare, energy management, real estate, politics, and education.

Blockchain technology is a revolutionary platform that has gained significant attention due to its potential to transform various industries. It is a decentralized and distributed ledger system that operates on a network of nodes, eliminating the need for a central authority or intermediary. Blockchain technology's key features and components include decentralization, immutability, transparency, security, smart contracts, distributed ledger, consensus mechanisms, anonymity and pseudonymity, interoperability, tokenization, scalability, and auditability (Baiod et al., 2021; Bodkhe et al., 2020).

Decentralization is a fundamental feature of blockchain technology that enhances security, transparency, and transaction trust. Immutability ensures that once data are recorded on the blockchain, it is extremely difficult to alter, creating a chain of blocks resistant to tampering (Politou et al., 2019). Transparency allows all participants in a blockchain network to have access to the same information, fostering trust among users as they can verify transactions and data independently. Security is achieved through advanced cryptographic techniques and consensus mechanisms, such as PoW or PoS, which add an extra layer of security (Akbar et al., 2021).

Smart contracts are self-executing contracts with the terms of the agreement written directly into code. They automatically enforce and execute the terms when predefined conditions are met, reducing the need for intermediaries (Mik, 2017). The distributed ledger is shared among all participants in the network, and each participant has a copy of the entire blockchain, enhancing resilience and eliminating a single point of failure. Consensus mechanisms ensure agreement among nodes and prevent double-spending (Akbar et al., 2021).

Anonymity and pseudonymity provide privacy and pseudonymity for users, while interoperability enables blockchain to facilitate interoperability between different systems and networks (Bernabe et al., 2019). Tokenization allows the creation of digital tokens representing ownership or access rights, and scalability is an ongoing challenge that research and development aim to address. Auditability provides an auditable trail of all activities on the network, which can be crucial for regulatory compliance and dispute resolution (Tremblay & Gendron, 2011).

The architecture of blockchain is divided into three main components: the network layer, the consensus layer, and the application layer. The network layer is responsible for communication between nodes on the network, while the consensus layer ensures that all nodes agree on the state of the blockchain (Xiao et al., 2020). Blockchain applications are developed and deployed in the application layer, including smart contracts, decentralized applications (DApps), and other blockchain-based services (Su et al., 2019).

Blockchain networks can be categorized into three main types: public, private, and consortium blockchains. Each type has unique features, benefits, and drawbacks, making them suitable for different use cases (Figure 4.1).

Public blockchains are open to everyone, allowing anyone to participate in the blockchain process, validate transactions, and access current records. They are decentralized and do not have any restrictions, making them ideal for applications that require high transparency and security. Public blockchains use consensus algorithms like PoW and PoS to verify transactions and add them to the ledger (Sheikh et al., 2018). No one owns them, and they are open to the public, making them highly distributed and secure. However, public blockchains can be slow due to their large size, and verifying each node can take time and effort. They also consume much energy due to the complex computations used to verify transactions (Potlapally et al., 2005).

Private or managed blockchains are closed networks where only authorized entities can validate transactions or data. They are used in networks where high privacy and security are required. Private blockchains are not open to the public, and access is restricted to authorized entities in the organization (Bernabe et al., 2019). They are less distributed than public blockchains, and a private authority develops the platform. Private blockchains are faster as they have some nodes for validations and offer customizability. They also provide strong privacy, as permission is needed to access transaction information. However, they are not truly decentralized as they require permission, and there is a risk of corruption as only a few participants are involved (Atzori, 2015).

Consortium blockchains are a hybrid of public and private blockchains. They are partially decentralized and allow a set of pre-selected nodes to validate transactions and data (Zheng et al., 2017). Consortium blockchains are used in networks with high privacy and security; collaboration between multiple organizations is needed. They are faster than public blockchains and provide strong privacy, as permission is needed to access transaction information. However, they are less distributed than public blockchains, and there is a risk of corruption as only a few participants are involved (Adam & Fazekas, 2021).

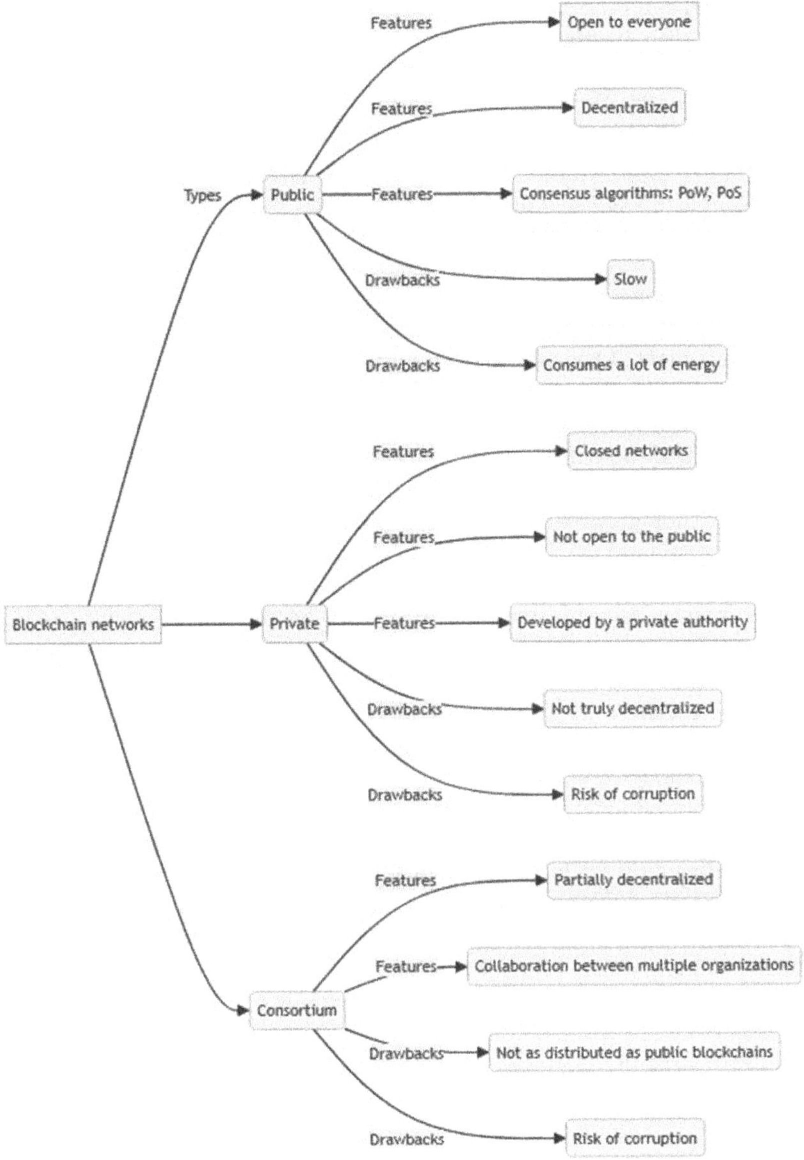

FIGURE 4.1   Types of blockchain networks and their features and drawbacks.

4.3   RENEWABLE ENERGY LANDSCAPE

The renewable energy landscape is transforming significantly, shifting from fossil fuels to affordable, renewable, and dependable energy sources. This shift is supported by the increasing cost-effectiveness of renewables, the potential economic advantages of clean energy employment, and the need to reduce dependence on fossil fuel imports (Berndes & Hansson, 2007).

Renewable energy sources include solar, wind, hydro, geothermal, and biomass. Solar and wind energy are the fastest-growing renewable energy sources, with solar PV remaining the powerhouse of growth in renewable electricity. Solar PV capacity additions are forecast to increase by 17% in 2021 to a new record of almost 160 GW, while onshore wind additions are set to be almost one-quarter higher on average than during the 2015–2020 period. Total offshore wind capacity is forecast to triple by 2026 (*Renewable electricity growth is accelerating faster than ever worldwide, supporting the emergence of the new global energy economy, 2021*).

Global trends in renewable energy adoption are positive, with the share of renewable energy in primary energy supply projected to grow from 16% in 2020 to 77% in 2050. The increasing cost-effectiveness of renewables drives this growth, the need to reduce greenhouse gas emissions, and the potential economic advantages of clean energy employment (Akaev & Davydova, 2023).

However, challenges remain in the renewable energy landscape, including access to finance and localized challenges, particularly in developing countries. A cumulative US$150 trillion is required to realize the 1.5°C target by 2050, averaging over US$5 trillion annually. Although global investment across all energy transition technologies reached a record high of US$1.3 trillion in 2022, annual investment must more than quadruple to remain on the 1.5°C pathway. Compared with the Planned Energy Scenario, an additional US$47 trillion in cumulative investment is required by 2050 to remain on the 1.5°C pathway (Akaev & Davydova, 2023).

The renewable energy sector faces several challenges that require innovative solutions to accelerate its growth and adoption. The first major challenge is cost competitiveness with traditional fossil fuel-based energy sources. While the costs of renewable technologies like solar and wind have decreased significantly, the initial investment required for installation and infrastructure remains a significant barrier to their widespread adoption (Luthra et al., 2015). This challenge can be addressed through continued research and development efforts, policy incentives, and market-based mechanisms that promote the adoption of renewable energy technologies (Gan et al., 2007).

The intermittent nature of renewable energy sources, such as solar and wind, also poses challenges in integrating them into the existing power grid. Fluctuations in energy supply due to weather conditions require advanced energy storage solutions and grid infrastructure upgrades to manage peak production periods and ensure a stable energy supply. Innovations in energy storage technologies, such as batteries, pumped hydro storage, and other emerging solutions, can help address these challenges and improve the reliability and flexibility of renewable energy systems (Koohi-Kamali et al., 2013).

Policy and regulatory hurdles are another significant obstacle for the renewable energy industry. A significant obstacle to the sector's growth is the need for consistent and supportive policy frameworks. Governments play a crucial role in driving investment and market growth through long-term renewable energy targets, incentives, subsidies, and favorable regulatory environments. Collaboration between governments, industry organizations, and other stakeholders is essential to develop and implement policies that support the transition to renewable energy sources (Martinez & Komendantova, 2020).

Technological advancements are also crucial to improving the efficiency and reliability of renewable energy technologies. Innovations in areas like solar panel design, wind turbine technology, and energy storage systems are crucial to address existing limitations and enhance the viability of renewable energy sources. Continued research and development efforts and public–private partnerships can help drive technological advancements and accelerate the transition to sustainable energy sources (Mallett, 2013).

Finally, the need for a skilled workforce and expertise in renewable energy technologies poses a challenge as the sector expands. There is a growing need for professionals with specialized knowledge in renewable energy technologies, grid integration, and policy development. Collaboration between governments, educational institutions, and industry organizations is essential to promote training programs and educational initiatives to support the renewable energy workforce (McCoy et al., 2012).

4.4 BLOCKCHAIN APPLICATIONS IN RENEWABLE ENERGY

Blockchain technology has emerged as a transformative solution in the renewable energy sector, particularly in peer-to-peer (P2P) energy trading platforms. These platforms leverage blockchain's decentralized and transparent nature to enable direct energy transactions between producers and consumers, bypassing the need for intermediaries and fostering a more efficient and sustainable energy ecosystem (Soto et al., 2021). One of the key applications of blockchain in renewable energy is the facilitation of P2P energy trading. By utilizing blockchain's distributed ledger system, P2P energy trading platforms allow individuals or entities to buy and sell excess energy directly with one another. This direct interaction empowers consumers to become prosumers, enabling them to generate, consume, and trade energy within a decentralized network.

Blockchain technology ensures the security and transparency of energy transactions in P2P trading platforms. The immutable nature of blockchain records all transactions securely and transparently, providing a reliable and tamper-proof system for tracking energy exchanges. This transparency builds trust among participants and eliminates the need for third-party verification, reducing transaction costs and enhancing the efficiency of energy trading (Yap et al., 2023). Moreover, P2P energy trading platforms supported by blockchain technology offer benefits such as increased flexibility, lower transaction costs, and enhanced grid stability. Participants can dynamically respond to supply and demand fluctuations in near-real time, optimizing energy distribution and consumption. Smart contracts, a feature of blockchain technology, automate energy trading processes, ensuring seamless and secure transactions between parties (Esmat et al., 2021).

Renewable Energy Certificates (RECs) are a type of Energy Attribute Certificate (EAC) that represent proof of the generation of one megawatt-hour (MWh) of electricity from renewable energy sources. They are tradable, non-tangible energy commodities that serve as a market-based solution to promote the development of renewable energy sources by providing financial incentives and an option for companies to meet their sustainability goals and regulatory requirements related to renewable energy use (Bugaeva et al., 2023).

RECs differ from carbon credits, tradable certificates, or permits that allow organizations to emit a certain amount of carbon emissions. Carbon credits enable businesses to offset carbon emissions by investing in certified climate action projects, such as afforestation, renewable energy projects, and energy efficiency initiatives (van der Gaast et al., 2018).

The value chain of RECs involves several key players, including project owners, issuers, REC registries, and buyers. Project owners generate renewable energy and apply for certification of their renewable energy production, selling the RECs to interested buyers. Issuers validate and verify the renewable energy production of the project owner and issue RECs, typically adhering to internationally recognized standards such as the International Renewable Energy Certificate (I-REC) standard. REC registries track and manage the issuance, transfer, and retirement of RECs, ensuring transparency and accountability. Buyers of RECs include companies, organizations, or individuals seeking to reduce their carbon footprint, meet emissions targets, or demonstrate corporate social responsibility by supporting the transition to a clean energy economy (Li et al., 2022).

RECs can be traded on a Book and Claim trading platform, such as the one that Book & Claim Limited operates in the United Kingdom. These platforms aim to simplify the trading of RECs and other sustainability certificates, allowing buyers to claim that their purchased energy was generated from an eligible renewable energy resource (Zuo, 2022).

4.5   BENEFITS OF INTEGRATING BLOCKCHAIN WITH RENEWABLE ENERGY

Integrating blockchain technology in the renewable energy sector offers numerous benefits, including enhanced security and data integrity, decentralization and resilience, and efficiency gains and cost reductions (Gawusu et al., 2022).

First, blockchain technology can significantly enhance security and data integrity in the renewable energy sector. As a shared and immutable ledger, blockchain simplifies the transaction recording process and ensures transparency, making it difficult for unauthorized parties to manipulate or alter data. This transparency and security are particularly important in the context of renewable energy certificates (RECs) and carbon credits, where blockchain can ensure transparency and verification of RECs, attracting investments in clean energy (Juszczyk & Shahzad, 2022).

Second, blockchain technology can foster decentralization and resilience in the renewable energy sector. By enabling peer-to-peer energy trading, blockchain allows renewable energy providers to sell excess power directly to customers without intermediaries. This decentralization can promote the utilization of renewable energy sources and foster technology adoption for other industry advancements, such as electric mobility and smart batteries for energy storage (Ahl et al., 2022).

Finally, blockchain technology can lead to efficiency gains and cost reductions in the renewable energy sector. By automating various processes involved in generating and transmitting renewable power, blockchain can help automate various processes, reduce transaction costs, and enable peer-to-peer energy exchanges. Additionally, blockchain-based smart contracts can automate energy transactions, enabling real-time monitoring, control, and optimization of energy flows within the grid, improving grid management, and facilitating the Integration of renewable energy sources into the existing energy infrastructure (Hasankhani et al., 2021).

4.6 CHALLENGES AND BARRIERS TO ADOPTION

Scalability issues are a significant challenge to adopting new technologies, including blockchain, in the renewable energy sector. Scalability refers to the ability of a system to handle increasing amounts of work or traffic without compromising its performance. In the context of blockchain, scalability is a critical issue due to the limited number of transactions that can be processed per second, leading to slow transaction speeds and high transaction fees (Khan et al., 2021).

The technological barrier of scalability is a significant challenge to adopting blockchain in the renewable energy sector. Integrating blockchain into the renewable energy sector requires a scalable system that can handle the large data generated by renewable energy sources, including solar, wind, and hydropower. The current blockchain systems, such as Bitcoin and Ethereum, are not designed to handle such large volumes of data, leading to slow transaction speeds and high transaction fees (Hu et al., 2019).

Various solutions have been proposed to overcome the technological barrier of scalability, including adopting a hybrid cloud approach, leveraging open-source software, using automation and self-services to enhance the user experience, and focusing on collaboration and scalability. A hybrid cloud approach allows for greater flexibility, scalability, and interoperability of different environments, enabling faster innovation and delivery of applications and services (Gundu et al., 2020). Leveraging open-source software can reduce the cost and risk of adopting new technology by providing access to a large community of testers and users who can contribute to the quality and innovation of the software (Setia et al., 2012).

Regulatory and compliance concerns are significant challenges to adopting blockchain technology in various industries, including the renewable energy sector. Blockchain in regulatory compliance can help streamline and alleviate the strain of regulation on financial institutions' IT systems, reducing compliance costs and improving accuracy and efficiency. Blockchain technology can provide an indisputable audit trail, including records of procedures and tasks undertaken for each client and documents shared, providing a single source of the truth through distributed ledger technology (DLT), but owned centrally by the governing entity (Erri Pradeep et al., 2021).

However, there are also challenges associated with blockchain use in regulatory compliance. Performance and security issues may pose considerable barriers to adoption, and blockchain technology is still considered more secure than a conventional database but requires more robust infrastructures for widespread adoption. Additionally, there are regulatory and compliance challenges related to jurisdiction, technology-neutral regulatory regime, governance and legal documentation, liability, and intellectual property (IP) (Andanda, 2019).

Jurisdiction issues arise due to blockchain's ability to cross jurisdictional boundaries, requiring careful consideration of relevant laws and contractual relationships among nodes. Technology-neutral regulatory regimes make it difficult to interpret how regulation should apply and which participants should be caught, requiring careful assessment of the nature and activities of a blockchain network and its participants. Governance and legal documentation are necessary to properly document the relationship between the network operator (if any) and its participants through legally enforceable contracts, establishing a clear and robust governance model concerning interactions among participants in the network. Liability must be carefully assessed and documented within each layer of network participation, including risks relating to security, confidentiality, regulation, taxation, data protection, immutability, automation, and decentralization (Perera et al., 2020). Intellectual property (IP) issues arise due to the need to share the underlying technology, including its software, for value to be gained, requiring careful consideration of the specific nature of the blockchain in question, including its purposes, subject matter, and the relationship between the blockchain participants (Belderbos et al., 2014).

Interoperability is a significant challenge in the blockchain industry, as different blockchain networks operate independently with unique rules, consensus mechanisms, and security protocols. This isolation limits the scope of interactions and hinders broader technological integration. Interoperability is the capability of different blockchain networks to exchange data and value seamlessly, connecting isolated blockchain networks and enabling them to interact and share resources (Anthony Jnr, 2024).

Interoperability is vital for the maturation and widespread adoption of blockchain technology, fostering a collaborative environment where diverse blockchain networks can communicate, leading to enhanced functionality, more significant innovation, and increased efficiency. It also addresses scalability concerns, allowing blockchains to handle more transactions and operations without compromising speed or security. Some of the use cases of interoperability include cross-chain transactions in finance, supply chain transparency, integrated decentralized finance (DeFi), inter-blockchain data sharing, and gaming and digital collectibles (Ray, 2023).

Cross-chain protocols, blockchain bridges, and interoperability platforms are solutions to achieve blockchain interoperability. Cross-chain protocols facilitate direct communication and transactions between independent blockchains, allowing them to recognize and interact with each other's data and assets. Blockchain bridges act as connectors between different blockchains, enabling seamless communication and data transfer across various platforms. Interoperability platforms provide a unified infrastructure for different blockchain networks to communicate, enabling cross-chain transactions and data sharing (Wang & Nixon, 2021).

Despite the potential benefits of interoperability, there are challenges to achieving seamless communication and data transfer between different blockchain networks. Technical complexity, security concerns, and regulatory challenges are some of the significant barriers to achieving blockchain interoperability. However, various projects and initiatives are working on solving the interoperability challenge, including Polkadot, Cosmos, and Interledger Protocol (ILP). These projects focus on enabling seamless communication and data exchange between blockchains, enhancing connectivity, scalability, and utility in the blockchain ecosystem (Vo et al., 2018).

4.7 REGULATORY FRAMEWORK AND POLICY CONSIDERATIONS

The current regulatory framework governing renewable energy encompasses a range of laws and policies that shape the sector's operations and development. One key aspect is the Renewable Energy Modernization Rule, which outlines the conduct of renewable energy activities within the existing regulatory framework. This rule guides compliance, standards, and best practices for renewable energy projects, ensuring they align with legal requirements and industry norms (Sutherland et al., 2015) (Figure 4.2).

Complementing this rule is the Renewable Energy Policy Manual, which delves into the legislative landscape surrounding renewable energy. This manual explores how legislation can address barriers to developing renewable technologies, offering insights into ownership structures, policy trends, and the evolving regulatory environment. By providing a comprehensive overview of renewable energy policies, this manual is a valuable resource for stakeholders navigating the complex regulatory landscape (Hicks & Ison, 2018).

Moreover, regulations, standards, and incentives at federal, state, and local levels play a pivotal role in shaping the renewable energy sector. These laws cover various aspects of renewable energy, including project structuring, real estate considerations, and financial incentives to foster investment in sustainable energy sources. By offering a mix of regulatory frameworks and incentives, policymakers aim to drive the

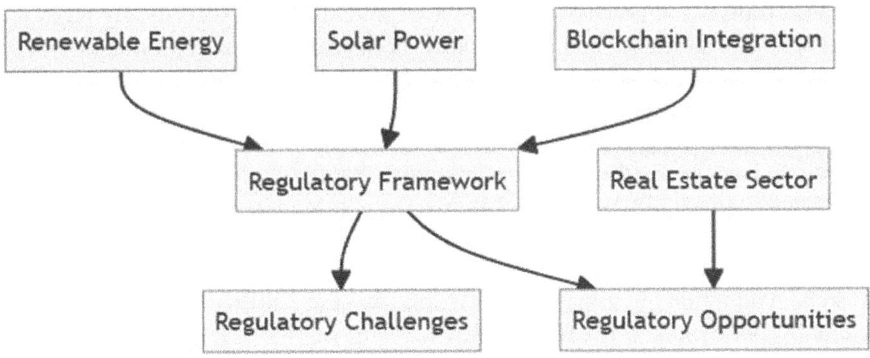

FIGURE 4.2 Key regulatory areas: renewable energy, solar power, blockchain, and real estate.

adoption of renewable energy technologies and accelerate the transition to a cleaner, more sustainable energy future (Vanegas Cantarero, 2020).

In the realm of solar power, specific policies and guidelines further shape the regulatory landscape. Documents such as Solar Power: Policy Overview and Good Practices outline renewable electricity standards, solar technology deployment strategies, and key considerations like interconnection requirements and tax incentives. These policies provide a roadmap for promoting solar energy adoption, ensuring alignment with broader energy goals and environmental objectives (Hernandez et al., 2020).

Additionally, regulatory frameworks and guidelines set by entities like the Bureau of Ocean Energy Management (BOEM) are crucial in overseeing renewable energy initiatives. BOEM adheres to environmental statutes, regulations, and executive orders that impact its work in areas such as offshore wind energy development. By complying with laws like the Oil Pollution Act, Energy Policy Act, and National Environmental Policy Act (NEPA), BOEM ensures that renewable energy projects are developed responsibly, balancing energy needs with environmental protection considerations (Russell et al., 2021).

Emerging policies to support blockchain integration are crucial for fostering the adoption and utilization of blockchain technology in various sectors. These policies aim to create a conducive environment for developing and implementing blockchain solutions. One key policy focus is knowledge sharing and developing new pilot projects to explore the potential applications of blockchain technology in different domains. By facilitating the exchange of information and best practices among stakeholders, policymakers aim to promote the adoption of blockchain technology and encourage innovation in various sectors (Pólvora et al., 2020).

Another emerging policy is the establishment of a common blockchain governance framework. This framework aims to set compliance conditions for validating nodes in permissioned blockchains, separating infrastructure and application layers to enhance the deployment speed, security, and efficiency of blockchain-based services in the public sector. By promoting permissioned blockchains, policymakers aim to streamline regulatory compliance and facilitate the deployment of blockchain use cases involving citizen data (Khan et al., 2021; Khan et al., 2022).

In addition, policymakers are increasingly supporting permissioned blockchain networks in the public sector. These networks offer advanced functionalities, increased transaction processing capabilities, and enhanced security. By promoting the adoption of permissioned blockchains, policymakers aim to stimulate innovation, encourage the development of new administrative processes, and harness the full potential of blockchain technology (Allen et al., 2020).

Furthermore, policymakers are incentivizing experimentation with blockchain technology and re-engineering administrative processes to leverage blockchain's transformative capabilities. By encouraging the exploration of innovative blockchain applications, policymakers aim to foster a culture of innovation and promote the development of new solutions that address the unique challenges associated with blockchain adoption in various sectors (Clohessy et al., 2019).

Finally, policymakers are increasingly focusing on initiatives that promote blockchain interoperability. Interoperability solutions aim to bridge different blockchain networks, enabling seamless data transfer and communication across disparate platforms. By supporting interoperability initiatives, policymakers seek to overcome the challenges associated with isolated blockchain domains and facilitate cross-chain transactions, thereby enhancing blockchain technology's overall utility and effectiveness (Bhat et al., 2022).

Regulatory challenges and opportunities surrounding blockchain technology are critical for its successful Integration into various sectors. One major challenge is the legal framework and territoriality of blockchains and shared distributed ledgers. Since these networks do not have a specific location, issues of jurisdiction, applicable law, and liability arise due to the lack of a central administration responsible for each ledger. This raises questions about establishing a legal framework to recognize blockchains as immutable and tamper-proof nodes, ensuring the veracity of information contained within them (Politou et al., 2019).

Another challenge is the recognition of blockchains as immutable sources of information. While there is consensus on the practical immutability of blocks, there needs to be more legal recognition of this aspect that hinders their use as trusted sources of identity and information. Smart contracts also present challenges, particularly in interpreting the "right to be forgotten" and determining liabilities associated with smart contract breaches or flaws in their design or execution (Herian, 2021).

Regulation for the Internet of Things (IoT) presents another challenge. Using blockchains as a valid regulatory registry for the IoT requires a legal framework that recognizes distributed ledgers as valid registries. This involves addressing territoriality, liability, and smart contract applicability challenges within the IoT context. However, there are also regulatory opportunities and solutions to these challenges. Governments can enact provisions tailored to smart contracts to clarify their legal status and reduce legal uncertainty. Regulatory sandboxes can be established to foster innovation in controlled environments, promoting the development and adoption of smart contract technologies (Alaassar et al., 2020).

International cooperation initiatives can also be crucial in addressing regulatory challenges associated with blockchain technology. By sharing best practices and coordinating regulatory approaches, countries can create a more harmonized and conducive environment for blockchain integration (Allen et al., 2020). The real estate sector is exploring using blockchain and smart contracts, known as "proptech", to streamline processes like land registration and title transfers. Applying blockchain technology in real estate presents opportunities for efficiency, transparency, and cost reduction in property transactions (Ullah & Al-Turjman, 2023).

4.8 FUTURE OUTLOOK AND EMERGING TRENDS

The future outlook and emerging trends in various industries are shaped by technological advancements, growth projections, market opportunities, and collaborations that are redefining the landscape. These trends drive innovation, competitiveness, and sustainability across sectors.

Technological advancements, such as Artificial Intelligence (AI) and Machine Learning (ML), are revolutionizing product modernization and driving product efficiency and intelligence. AI and ML technologies are enhancing existing features and pioneering novel capabilities that redefine product design (Verganti et al., 2020).

Additionally, sustainability has become a global demand, leading to eco-friendly design approaches and a focus on minimal environmental impact throughout the product lifecycle. These advancements are reshaping industries, driving innovation, and meeting the evolving needs of consumers (Abedsoltan, 2024).

The evolving technological landscape presents significant growth projections and market opportunities for businesses that embrace innovation. Companies investing in research and development, embracing new technologies, and fostering a culture of continuous innovation are well-positioned to capitalize on emerging trends. Integrating AI, sustainability practices, IoT proliferation, personalization, and agile development is shaping products and defining the future of business and consumer interaction. Understanding and implementing these trends is essential for staying competitive and meeting consumer demands in a rapidly evolving market (Mason-Jones & Towill, 1997).

Collaboration and partnerships play a crucial role in shaping industries and driving innovation. Organizations collaborating with technology partners, industry experts, and research institutions can leverage collective expertise to navigate regulatory challenges, drive technological advancements, and capitalize on market opportunities. By fostering a culture of collaboration, businesses can stay agile, adapt to changing market dynamics, and create synergies that drive industry growth and transformation (Camarinha-Matos et al., 2019).

4.9 SUMMARY

The exploration of blockchain technology's integration with renewable energy sources reveals a transformative potential for the energy sector. Blockchain offers unique capabilities that can accelerate the adoption of renewable energy by addressing key challenges such as decentralization, energy independence, and transparency. Blockchain enables streamlined energy trading, peer-to-peer transactions, and grid optimization through tokenizing renewable energy assets and smart contracts. These advancements enhance efficiency and promote sustainability and decentralization within the energy ecosystem.

Stakeholders in the renewable energy sector, including policymakers, businesses, and individuals, are urged to embrace the opportunities presented by blockchain technology. To harness the full potential of blockchain for renewable energy, stakeholders should focus on the following actions:

First, clear regulatory frameworks that support blockchain technology integration in the energy sector should be developed. Address scalability, interoperability, and data privacy challenges to ensure a conducive environment for innovation and adoption. Second, foster collaboration among utilities, regulators, technology providers, and consumers to establish standardized protocols and seamless integration of blockchain solutions. Collaborative efforts are essential for overcoming barriers and driving widespread adoption. Third, encourage innovation and investment in blockchain-based solutions for energy trading, grid management, and renewable energy production. Embrace emerging technologies to enhance the energy sector's transparency, efficiency, and sustainability. Fourth, raise awareness about the benefits of blockchain technology in renewable energy and educate stakeholders about its potential applications. Promote knowledge-sharing initiatives to empower stakeholders to make informed decisions and drive positive change. Lastly, emphasize the role of blockchain in promoting sustainability, reducing carbon emissions, and accelerating the transition to a decentralized energy system. Encourage the adoption of renewable energy sources and energy-efficient practices to build a more resilient and sustainable energy ecosystem.

By taking proactive steps to leverage blockchain technology, stakeholders in the renewable energy sector can unlock new business models, empower consumers, and drive the transition toward a more sustainable and decentralized energy system. Embracing innovation and collaboration is key to realizing the full potential of blockchain in revolutionizing the future of renewable energy.

REFERENCES

Abedsoltan, H. (2024). Applications of plastics in the automotive industry: Current trends and future perspectives. *Polymer Engineering & Science, 64*(3), 929–950. https://doi.org/10.1002/pen.26604

Adam, I., & Fazekas, M. (2021). Are emerging technologies helping win the fight against corruption? A review of the state of evidence. *Information Economics and Policy, 57*, 100950.

Ahl, A., Goto, M., Yarime, M., Tanaka, K., & Sagawa, D. (2022). Challenges and opportunities of blockchain energy applications: Interrelatedness among technological, economic, social, environmental, and institutional dimensions. *Renewable and Sustainable Energy Reviews, 166*, 112623. https://doi.org/10.1016/j.rser.2022.112623

Akaev, A., & Davydova, O. (2023). Climate and energy: Energy transition scenarios and global temperature changes based on current technologies and trends. *Reconsidering the Limits to Growth: A Report to the Russian Association of the Club of Rome* (pp. 53–70). Springer.

Akbar, N. A., Muneer, A., ElHakim, N., & Fati, S. M. (2021). Distributed hybrid double-spending attack prevention mechanism for proof-of-work and proof-of-stake blockchain consensuses. *Future Internet, 13*(11), 285.

Alaassar, A., Mention, A.-L., & Aas, T. H. (2020). Exploring how social interactions influence regulators and innovators: The case of regulatory sandboxes. *Technological Forecasting and Social Change, 160*, 120257. https://doi.org/10.1016/j.techfore.2020.120257

Allen, D. W. E., Berg, C., Markey-Towler, B., Novak, M., & Potts, J. (2020). Blockchain and the evolution of institutional technologies: Implications for innovation policy. *Research Policy, 49*(1), 103865. https://doi.org/10.1016/j.respol.2019.103865

Andanda, P. (2019). Towards a paradigm shift in governing data access and related intellectual property rights in big data and health-related research. *IIC – International Review of Intellectual Property and Competition Law, 50*(9), 1052–1081. https://doi.org/10.1007/s40319-019-00873-2

Andoni, M., Robu, V., Flynn, D., Abram, S., Geach, D., Jenkins, D., McCallum, P., & Peacock, A. (2019). Blockchain technology in the energy sector: A systematic review of challenges and opportunities. *Renewable and Sustainable Energy Reviews, 100*, 143–174.

Anees, A. S. (2012). Grid integration of renewable energy sources: Challenges, issues and possible solutions. *2012 IEEE 5th India International Conference on Power Electronics (IICPE)*.

Anthony Jnr, B. (2024). Enhancing blockchain interoperability and intraoperability capabilities in collaborative enterprise-a standardized architecture perspective. *Enterprise Information Systems, 18*(3), 2296647. https://doi.org/10.1080/17517575.2023.2296647

Asim, M. (2017). A survey on application layer protocols for internet of things (IoT). *International Journal of Advanced Research in Computer Science, 8*(3).

Atzori, M. (2015). Blockchain technology and decentralized governance: Is the state still necessary? Available at: *SSRN 2709713.*

Baashar, Y., Alkawsi, G., Alkahtani, A. A., Hashim, W., Razali, R. A., & Tiong, S. K. (2021). Toward blockchain technology in the energy environment. *Sustainability, 13*(16), 9008.

Baiod, W., Light, J., & Mahanti, A. (2021). Blockchain technology and its applications across multiple domains: A survey. *Journal of International Technology and Information Management, 29*(4), 78–119.

Belderbos, R., Cassiman, B., Faems, D., Leten, B., & Van Looy, B. (2014). Co-ownership of intellectual property: Exploring the value-appropriation and value-creation implications of co-patenting with different partners. *Research Policy, 43*(5), 841–852. https://doi.org/10.1016/j.respol.2013.08.013

Bernabe, J. B., Canovas, J. L., Hernandez-Ramos, J. L., Moreno, R. T., & Skarmeta, A. (2019). Privacy-preserving solutions for blockchain: Review and challenges. *IEEE Access, 7,* 164908–164940.

Berndes, G., & Hansson, J. (2007). Bioenergy expansion in the EU: Cost-effective climate change mitigation, employment creation and reduced dependency on imported fuels. *Energy Policy, 35*(12), 5965–5979.

Bhat, S. A., Huang, N.-F., Sofi, I. B., & Sultan, M. (2022). Agriculture-food supply chain management based on blockchain and IoT: A narrative on enterprise blockchain interoperability. *Agriculture, 12*(1), 40. www.mdpi.com/2077-0472/12/1/40

Bodkhe, U., Tanwar, S., Parekh, K., Khanpara, P., Tyagi, S., Kumar, N., & Alazab, M. (2020). Blockchain for Industry 4.0: A comprehensive review. *IEEE Access, 8,* 79764–79800.

Bugaeva, T., Grishacheva, A., & Novikova, O. (2023). Development of tools for decarbonization of electricity consumption in the Russian federation. In *Digital Transformation on Manufacturing, Infrastructure & Service*, Cham.

Camarinha-Matos, L. M., Fornasiero, R., Ramezani, J., & Ferrada, F. (2019). Collaborative Networks: A pillar of digital transformation. *Applied Sciences, 9*(24), 5431. www.mdpi.com/2076-3417/9/24/5431

Clohessy, T., Acton, T., & Rogers, N. (2019). Blockchain adoption: Technological, organisational and environmental considerations. In H. Treiblmaier & R. Beck (Eds.), *Business Transformation through Blockchain: Volume I* (pp. 47–76). Springer International Publishing. https://doi.org/10.1007/978-3-319-98911-2_2

Erri Pradeep, A. S., Yiu, T. W., Zou, Y., & Amor, R. (2021). Blockchain-aided information exchange records for design liability control and improved security. *Automation in Construction, 126,* 103667. https://doi.org/10.1016/j.autcon.2021.103667

Esmat, A., de Vos, M., Ghiassi-Farrokhfal, Y., Palensky, P., & Epema, D. (2021). A novel decentralized platform for peer-to-peer energy trading market with blockchain technology. *Applied Energy, 282,* 116123. https://doi.org/10.1016/j.apenergy.2020.116123

Gan, L., Eskeland, G. S., & Kolshus, H. H. (2007). Green electricity market development: Lessons from Europe and the US. *Energy Policy, 35*(1), 144–155.

Gawusu, S., Zhang, X., Ahmed, A., Jamatutu, S. A., Miensah, E. D., Amadu, A. A., & Osei, F. A. J. (2022). Renewable energy sources from the perspective of blockchain integration: From theory to application. *Sustainable Energy Technologies and Assessments, 52,* 102108. https://doi.org/10.1016/j.seta.2022.102108

Gundu, S. R., Panem, C. A., & Thimmapuram, A. (2020). Hybrid IT and multi cloud an emerging trend and improved performance in cloud computing. *SN Computer Science, 1*(5), 256. https://doi.org/10.1007/s42979-020-00277-x

Hasankhani, A., Mehdi Hakimi, S., Bisheh-Niasar, M., Shafie-khah, M., & Asadolahi, H. (2021). Blockchain technology in the future smart grids: A comprehensive review and frameworks. *International Journal of Electrical Power & Energy Systems, 129*, 106811. https://doi.org/10.1016/j.ijepes.2021.106811

Herian, R. (2021). Smart contracts: A remedial analysis. *Information & Communications Technology Law, 30*(1), 17–34. https://doi.org/10.1080/13600834.2020.1807134

Hernandez, R. R., Jordaan, S. M., Kaldunski, B., & Kumar, N. (2020). Aligning climate change and sustainable development goals with an innovation systems roadmap for renewable power [original research]. *Frontiers in Sustainability, 1.* https://doi.org/10.3389/frsus.2020.583090

Hicks, J., & Ison, N. (2018). An exploration of the boundaries of 'community' in community renewable energy projects: Navigating between motivations and context. *Energy Policy, 113*, 523–534. https://doi.org/10.1016/j.enpol.2017.10.031

Hu, Y., Manzoor, A., Ekparinya, P., Liyanage, M., Thilakarathna, K., Jourjon, G., & Seneviratne, A. (2019). A delay-tolerant payment scheme based on the ethereum blockchain. *IEEE Access, 7*, 33159–33172. https://doi.org/10.1109/ACCESS.2019.2903271

Javaid, M., Haleem, A., Singh, R. P., Khan, S., & Suman, R. (2021). Blockchain technology applications for Industry 4.0: A literature-based review. *Blockchain: Research and Applications, 2*(4), 100027.

Juszczyk, O., & Shahzad, K. (2022). Blockchain technology for renewable energy: Principles, applications and prospects. *Energies, 15*(13), 4603. www.mdpi.com/1996-1073/15/13/4603

Khan, D., Jung, L. T., & Hashmani, M. A. (2021). Systematic literature review of challenges in blockchain scalability. *Applied Sciences, 11*(20), 9372. www.mdpi.com/2076-3417/11/20/9372

Khan, S., Shael, M., Majdalawieh, M., Nizamuddin, N., & Nicho, M. (2022). Blockchain for governments: The case of the Dubai government. *Sustainability, 14*(11), 6576. www.mdpi.com/2071-1050/14/11/6576

Koohi-Kamali, S., Tyagi, V., Rahim, N., Panwar, N., & Mokhlis, H. (2013). Emergence of energy storage technologies as the solution for reliable operation of smart power systems: A review. *Renewable and Sustainable Energy Reviews, 25*, 135–165.

Li, P., Ng, J., & Lu, Y. (2022). Accelerating the adoption of renewable energy certificate: Insights from a survey of corporate renewable procurement in Singapore. *Renewable Energy, 199*, 1272–1282. https://doi.org/10.1016/j.renene.2022.09.066

Luthra, S., Kumar, S., Garg, D., & Haleem, A. (2015). Barriers to renewable/sustainable energy technologies adoption: Indian perspective. *Renewable and Sustainable Energy Reviews, 41*, 762–776.

Mallett, A. (2013). Technology cooperation for sustainable energy: A review of pathways. *Wiley Interdisciplinary Reviews: Energy and Environment, 2*(2), 234–250.

Martinez, N., & Komendantova, N. (2020). The effectiveness of the social impact assessment (SIA) in energy transition management: Stakeholders' insights from renewable energy projects in Mexico. *Energy Policy, 145*, 111744.

Mason-Jones, R., & Towill, D. R. (1997). Information enrichment: Designing the supply chain for competitive advantage. *Supply Chain Management: An International Journal, 2*(4), 137–148. https://doi.org/10.1108/13598549710191304

McCoy, A. P., O'Brien, P., Novak, V., & Cavell, M. (2012). Toward understanding roles for education and training in improving green jobs skills development. *International Journal of Construction Education and Research, 8*(3), 186–203.

Mik, E. (2017). Smart contracts: Terminology, technical limitations and real world complexity. *Law, Innovation and Technology, 9*(2), 269–300.

Perera, S., Nanayakkara, S., Rodrigo, M. N. N., Senaratne, S., & Weinand, R. (2020). Blockchain technology: Is it hype or real in the construction industry? *Journal of Industrial Information Integration, 17,* 100125. https://doi.org/10.1016/j.jii.2020.100125

Pfeifer, A., Dobravec, V., Pavlinek, L., Krajačić, G., & Duić, N. (2018). Integration of renewable energy and demand response technologies in interconnected energy systems. *Energy, 161,* 447–455.

Politou, E., Casino, F., Alepis, E., & Patsakis, C. (2019). Blockchain mutability: Challenges and proposed solutions. *IEEE Transactions on Emerging Topics in Computing, 9*(4), 1972–1986.

Pólvora, A., Nascimento, S., Lourenço, J. S., & Scapolo, F. (2020). Blockchain for industrial transformations: A forward-looking approach with multi-stakeholder engagement for policy advice. *Technological Forecasting and Social Change, 157,* 120091. https://doi.org/10.1016/j.techfore.2020.120091

Potlapally, N. R., Ravi, S., Raghunathan, A., & Jha, N. K. (2005). A study of the energy consumption characteristics of cryptographic algorithms and security protocols. *IEEE Transactions on Mobile Computing, 5*(2), 128–143.

Ray, P. P. (2023). Web3: A comprehensive review on background, technologies, applications, zero-trust architectures, challenges and future directions. *Internet of Things and Cyber-Physical Systems, 3,* 213–248. https://doi.org/10.1016/j.iotcps.2023.05.003

Renewable electricity growth is accelerating faster than ever worldwide, supporting the emergence of the new global energy economy. (2021). www.iea.org/news/renewable-electricity-growth-is-accelerating-faster-than-ever-worldwide-supporting-the-emergence-of-the-new-global-energy-economy/

Russell, A., Bingaman, S., & Garcia, H.-M. (2021). Threading a moving needle: The spatial dimensions characterizing US offshore wind policy drivers. *Energy Policy, 157,* 112516. https://doi.org/10.1016/j.enpol.2021.112516

Setia, P., Rajagopalan, B., Sambamurthy, V., & Calantone, R. (2012). How peripheral developers contribute to open-source software development. *Information Systems Research, 23*(1), 144–163. https://doi.org/10.1287/isre.1100.0311

Sheikh, H., Azmathullah, R. M., & Rizwan, F. (2018). Proof-of-work vs proof-of-stake: A comparative analysis and an approach to blockchain consensus mechanism. *International Journal for Research in Applied Science & Engineering Technology, 6*(12), 786–791.

Soto, E. A., Bosman, L. B., Wollega, E., & Leon-Salas, W. D. (2021). Peer-to-peer energy trading: A review of the literature. *Applied Energy, 283,* 116268.

Su, C., Liu, Y., Li, R., Wu, W., Fawcett, J. P., & Gu, J. (2019). Absorption, distribution, metabolism and excretion of the biomaterials used in nanocarrier drug delivery systems. *Advanced Drug Delivery Reviews, 143,* 97–114.

Sutherland, L.-A., Peter, S., & Zagata, L. (2015). Conceptualising multi-regime interactions: The role of the agriculture sector in renewable energy transitions. *Research Policy, 44*(8), 1543–1554. https://doi.org/10.1016/j.respol.2015.05.013

Tremblay, M.-S., & Gendron, Y. (2011). Governance prescriptions under trial: On the interplay between the logics of resistance and compliance in audit committees. *Critical Perspectives on Accounting, 22*(3), 259–272.

Ullah, F., & Al-Turjman, F. (2023). A conceptual framework for blockchain smart contract adoption to manage real estate deals in smart cities. *Neural Computing and Applications, 35*(7), 5033–5054. https://doi.org/10.1007/s00521-021-05800-6

van der Gaast, W., Sikkema, R., & Vohrer, M. (2018). The contribution of forest carbon credit projects to addressing the climate change challenge. *Climate Policy, 18*(1), 42–48. https://doi.org/10.1080/14693062.2016.1242056

Vanegas Cantarero, M. M. (2020). Of renewable energy, energy democracy, and sustainable development: A roadmap to accelerate the energy transition in developing countries. *Energy Research & Social Science, 70,* 101716. https://doi.org/10.1016/j.erss.2020.101716

Verganti, R., Vendraminelli, L., & Iansiti, M. (2020). Innovation and design in the age of artificial intelligence. *Journal of Product Innovation Management, 37*(3), 212–227. https://doi.org/10.1111/jpim.12523

Vo, H. T., Wang, Z., Karunamoorthy, D., Wagner, J., Abebe, E., & Mohania, M. (30 Jul.–3 Aug. 2018). Internet of blockchains: Techniques and challenges ahead. *2018 IEEE International Conference on Internet of Things (iThings) and IEEE Green Computing and Communications (GreenCom) and IEEE Cyber, Physical and Social Computing (CPSCom) and IEEE Smart Data (SmartData).*

Wang, G., & Nixon, M. (6–8 Dec. 2021). InterTrust: Towards an efficient blockchain interoperability architecture with trusted services. *2021 IEEE International Conference on Blockchain (Blockchain).*

Wang, Q., & Su, M. (2020). Integrating blockchain technology into the energy sector – from theory of blockchain to research and application of energy blockchain. *Computer Science Review, 37,* 100275.

Wu, K., Ma, Y., Huang, G., & Liu, X. (2021). A first look at blockchain-based decentralized applications. *Software: Practice and Experience, 51*(10), 2033–2050.

Xiao, Y., Zhang, N., Lou, W., & Hou, Y. T. (2020). A survey of distributed consensus protocols for blockchain networks. *IEEE Communications Surveys & Tutorials, 22*(2), 1432–1465.

Yang, F., Zhou, W., Wu, Q., Long, R., Xiong, N. N., & Zhou, M. (2019). Delegated proof of stake with downgrade: A secure and efficient blockchain consensus algorithm with downgrade mechanism. *IEEE Access, 7,* 118541–118555.

Yap, K. Y., Chin, H. H., & Klemeš, J. J. (2023). Blockchain technology for distributed generation: A review of current development, challenges and future prospect. *Renewable and Sustainable Energy Reviews, 175,* 113170. https://doi.org/10.1016/j.rser.2023.113170

Zheng, Z., Xie, S., Dai, H., Chen, X., & Wang, H. (2017). An overview of blockchain technology: Architecture, consensus, and future trends. *2017 IEEE International Congress on Big Data (BigData Congress).*

Zuo, Y. (2022). Tokenizing Renewable Energy Certificates (RECs) – a blockchain approach for REC issuance and trading. *IEEE Access, 10,* 134477–134490. https://doi.org/10.1109/ACCESS.2022.3230937

5 Sustainable Agriculture and Food Supply Chains

Mojtaba Rezaie and Aydin Shishegaran

5.1 INTRODUCTION

Sustainable agriculture and food supply chains encompass the entire food journey, from farming practices to reaching the consumers. The main objective is ensuring this journey benefits people, the economy, and farmers (Rejeb et al., 2022). Because some factors can threaten food supply chains and agriculture practices, like water pollution, greenhouse gas emissions, etc., economic disadvantages for farmers or unconditional farming practices can significantly reduce productivity in food production systems (Rehman et al., 2022).

Therefore, it is critically recommended that an efficient supply chain system to minimize environmental impact be provided to invite healthy and sustainable agriculture and food supply chains. Building a sustainable food system is an important mission to implement this picture, and it is necessary to promote a secure food supply for future generations (Irani & Sharif, 2016).

The next section focuses on challenges and issues in agriculture and food supply chains and demonstrates the importance of risk factors in sustainable agriculture production.

5.2 CHALLENGES AND CRITICAL ISSUES

Sustainability issues are especially pressing in developing countries. With a rapidly growing population, these nations face immense pressure on their food systems (Cicciù et al., 2022). For example, as a major agricultural producer and consumer, India exemplifies this challenge. Despite being a top producer of fruits, vegetables, and other food sources, a significant portion of this produce goes to waste due to inefficiencies in the food supply chain. This and social issues like malnutrition and poverty highlight India's urgent need for a more sustainable food system (Sharma et al., 2023).

Despite significant challenges, promising solutions are emerging. For example, circular economy (CE) offers a model for recycling existing materials and creating a pathway to sustainable consumption. This is a framework to reduce waste, spoilage, and resource overuse in food supply chains. Likewise, it is compatible with sustainable development goals (SDGs). CE's substantial merit is in turning waste into valuable resources that can be fed into the food system.

Technology advancements like Industry 4.0 (I4.0) can contribute to a more sustainable future. I4.0 encompasses AI, AR, and IoT technologies, creating a smart and sustainable ecosystem. These ecosystems reduce waste and optimize the supply chain

DOI: 10.1201/9781003609865-5

through data analytics like the novel growing start-ups such as "Lithos Carbon", "Farmspeak Technology", and "AeroFarms". The technologies mentioned can connect people, processes, and organizations, allowing for better resource management and efficiency. By employing data and technology applications, accessible resources are managed effectively, and processes improve efficiently. This highlights opportunities for a sustainable food system (Kumar et al., 2024) and provides a healthy constructed supply chain for future generations.

Data analysis is key to unlocking a more circular and sustainable food supply chain (Kumar et al., 2022). Big data can be used to analyze current waste streams and identify areas for improvement. It can also be used to understand consumer preferences, allowing for better production planning and reduced waste. For instance, McCormick's "flavor forecast" uses big data to predict customer trends and optimize product development (Tao et al., 2021).

Blockchain technology offers another promising solution for a more sustainable future. Creating a transparent and secure record of transactions can build trust within the food supply chain (Chandan et al., 2023). It can track land use and product origin (provenance) and manage financial transactions. This allows for quicker identification of contamination sources and better product recall management.

Beyond big data and blockchain, other Industry 4.0 technologies are crucial. Sensor technology and cloud computing provide real-time monitoring of perishable goods. This allows for better decision-making and helps reduce waste due to spoilage or improper handling. Artificial intelligence (AI) can also optimize logistics and route planning, increasing efficiency (Subramanian et al., 2020).

Therefore, Industry 4.0 technologies hold immense potential for a more sustainable food supply chain. However, strategic planning and innovation are crucial to ensure their successful adoption and maximize the benefits for global food systems.

5.3 BARRIERS TO I4.0 ADOPTION IN SUSTAINABLE FOOD SUPPLY CHAINS (SFSCS)

In developing countries' business environments, significant barriers exist to adopting I4.0 in SFSCs. Resistance to change from established business models and a lack of skilled personnel with the technical expertise for I4.0 implementation can hinder progress (Kumar et al., 2024).

The intricate nature of existing supply chains in developing countries can be a significant barrier to I4.0 adoption in SFSCs. These complex networks, often characterized by numerous stakeholders, fragmented operations, and limited data sharing, can make it challenging to integrate the interconnected systems and real-time data analysis that I4.0 relies on. While I4.0 has the potential to streamline these complexities, overcoming them initially can be an obstacle for SFSCs looking to implement these new technologies (Ozkan-Ozen et al., 2020).

5.4 SWOT APPROACH TO FOOD MANUFACTURING

By addressing the weaknesses and capitalizing on the opportunities, the food manufacturer can strengthen its position in the market. The below SWOT analysis is a proactive measure to mitigate threats that will ensure long-term business success in the dynamic food industry and help shape sustainable food manufacturing (Dania et al., 2024).

As Figure 5.1 indicates, high-quality products and competitiveness in the price of the products are strengths, and exporting products is an opportunity. Product sales through local and global markets will drive sales growth and, consequently, the products. Retailers have a significant role in increasing sales and are good partners as local sellers. These are those who can make the brand popular and usable. Branding is the essence of consumer demand, so working on branding in business is a game-changer strategy.

In addition to that, accessibility and reachability of raw materials is an environmental issue; it also is a factor that can boost sales. Competitiveness in producing innovative goods depends on the mentioned factors. Collaboration with stakeholders and strong partnerships between the private and public sectors provide standardized processes for having secure and sustainable production sources. This means proactively standing on competitive threats.

The workforce limitations and operational inefficiencies are addressing a big challenge. Good governance, governmental support, and flexible work arrangements can respond to work needs situations. Operations should be integrated into digital capabilities. Digitalized operation hints at accurate and optimal processes, and more importantly, it helps to achieve innovative preventive maintenance processes. This leads to adopting digital tools that automate the production process.

Due to the common ground of human errors in the food manufacturing process, it is necessary to automate some error-prone tasks, like manually handling financial records and, generally, repeatable and boring processes related to human intervention. In workflow, compliance with e-commerce principles and advertising campaigns in social media are effective ways of increasing productivity as practical strategies.

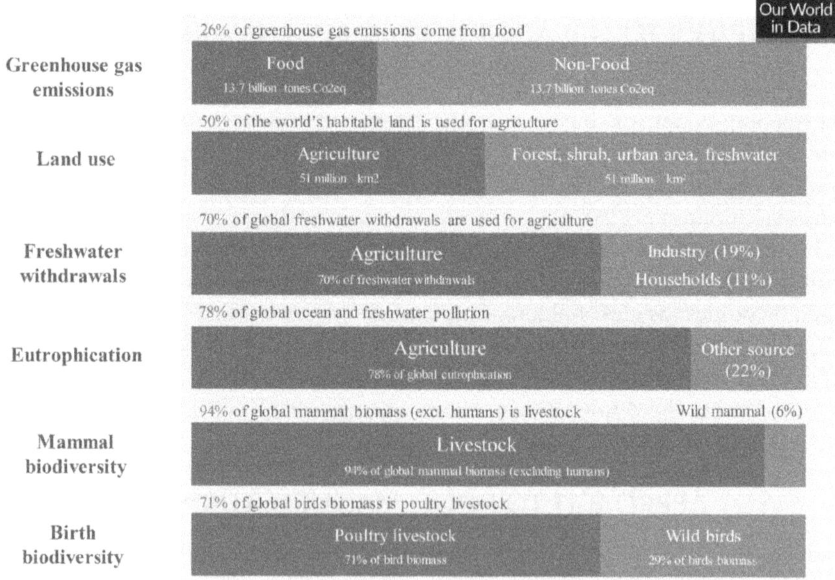

FIGURE 5.1 SWOT approach to food manufacturing.

5.5 OPTIMIZING RESOURCE USE AND MINIMIZING EMISSIONS WITHIN FOOD SUPPLY CHAINS (FSCS)

This pointed to a set of strategies aimed at reducing the environmental footprint of the journey food, which consists of reducing food waste, water conservation, sustainable land management, etc. (Krishnan et al., 2020).

The rising consumer demand for locally sourced food leads to these titled strategies. The mentioned strategic trend aligns perfectly with the growing awareness of environmental responsibility. Short Food Supply Chains (SFSCs) have traditionally been championed for their focus on social and institutional proximity. This means fostering closer connections between producers, distributors, and consumers within a defined local area (Reina-Usuga et al., 2020).

However, the emphasis on geographical closeness in SFSCs does not necessarily equate to environmental sustainability. While shorter distances might translate to reduced transportation emissions, other factors come into play. Localized production methods might only sometimes be the most efficient regarding resource utilization, and potential limitations in technology or expertise within a local area could hinder overall environmental impact reduction.

5.6 SHORT FOOD SUPPLY CHAINS

Food waste and loss continue to rise, even with industry efforts to reduce them. Fresh produce, especially fruits and vegetables, are particularly prone to spoilage due to their short shelf life and sensitivity to various conditions. Inefficiencies and logistical hurdles throughout the production chain also contribute significantly to these losses. Connecting producers directly to consumers might help reduce waste, but its impact on environmental impact is not straightforward. Factors like transportation methods, production practices, and supply chain design all play a role (Bayir et al., 2022).

SFSCs, as defined by the European Union, involve small businesses working together to support local economies and foster close relationships between producers, processors, and consumers. These systems can benefit producers with increased profits and offer consumers greater transparency about the origin of their food (Silva et al., 2024).

The idea of SFSCs being inherently good for the environment is being challenged. While they often offer social benefits and potentially higher profits, their environmental impact is not only sometimes positive (Malak-Rawlikowska et al., 2019). Some studies suggest shorter chains might even have higher emissions per unit of product (Gonçalves & Zeroual, 2017; Mancini et al., 2019).

However, the picture is more complex. Other research highlights the nuanced nature of environmental sustainability in SFSCs (Vitters\u00f8 et al., 2019). Recent frameworks even propose that short and long chains can co-exist and interact in various ways, challenging the idea that they are strictly competing models. These interactions can be independent, competitive, cooperative, or even coordinated.

Furthermore, studies show that short and long food chains can demonstrate resilience during disruptions like severe weather events. This highlights the importance of considering scale, diversity, flexibility, and collaboration when evaluating sustainability (Thamarai et al., 2024).

5.7 THE RISE OF FOOD PROCESSING TECHNOLOGIES

Consumers often place greater trust in food sourced locally (Food, 2021), which fuels the growth of SFSCs. However, scaling down processing technologies for these smaller chains can take time and effort. Large-scale food chains are optimized for high-volume production, and cost-effective technologies like thermal processing may not be easily adapted for smaller operations due to upfront costs, space limitations, and reliable electricity access (Lee et al., 2021).

Fortunately, new technologies are emerging that can address this challenge (Moller et al., 2023). High-pressure processing (HPP), pulsed electric field (PEF), and ultrasound are promising alternatives that can effectively process smaller volumes of food. These newer methods offer advantages like lower initial investment, reduced operating costs, faster processing times, and the ability to adapt to different production scales. Additionally, the consumer preference for minimally processed foods further drives the adoption of these emerging technologies (Alexandrova-Stefanova et al., 2023).

The mentioned technologies are gaining significant traction in the food industry. Since 2017 (Aganovic et al., 2017), these processing units have more than doubled worldwide. However, traditional processing methods sometimes offer greater efficiency despite those methods imposing potentially higher operational costs (Silva et al., 2024).

5.8 MERITS OF BLOCKCHAIN AND DECENTRALIZED PROCESSING

Due to the previous considerations about technology application in food supply chains, more merits exist in adopting technologies, especially for shorter food supply chains. For example, decentralized processing (Ge et al., 2017) allows for more local sourcing of ingredients, strengthening local economies. In the following, agricultural ecosystems will face growth with the aim of security and sustainable food production. Blockchain and decentralized processing increase the overall resilience and adaptability of the food system, and according to the vulnerability of shorter chains, it is quite obvious that blockchain increases transaction transparency, food safety, and security and also reduces food fraud and waste.

The blockchain application in agriculture describes how data are stored securely across a network, the process that makes it work, the tools it uses, the network itself, or a new way to manage economic transactions (Leng et al., 2018). This technology's characteristics are accountability, transparency, flexibility, availability, usability, manageability, and sustainability (Krithika, 2022), which direct food systems toward sustainable production and secure chains.

Blockchain technology offers a powerful tool for managing agricultural data, particularly land use, pesticide application, product traceability, and financial flows (Kamble et al., 2020). This technology goes beyond simple disruption, potentially forming the bedrock of new economic and social systems. In sustainable agriculture, blockchain provides the infrastructure for digitization, automation, and real-time tracking, all crucial aspects of driving farmer-led sustainability initiatives. According to consumers' demands for more information about the food they buy, making trust in the agricultural food supply chain (AFSC) will be the key to success for producers of products. This subject reflects the importance of transparency in producing goods. Digital product passports, digital eco-labeling, and detailed supplier certification are some practices that can help digitally gain product transparency. These practices are against malicious activities such as fraud and food security quality threats. Digital tools such as blockchain and its solutions can provide consumers with reliable and verifiable information about the origin of food (Ge et al., 2017).

5.9 RENEWABLE ENERGY CONSUMPTION (RECS)

Renewable energy sources are energy sources that can be naturally replenished on a human timescale. These sources contain solar energy, wind, geothermal, hydropower, bioenergy, etc. The International Energy Agency (IEA) reported in 2022 that renewable energy supply had increased by close to 8%, indicating that consumption of renewable energies has increased over past years. In sustainable agriculture, **RECs** play a key role by reducing reliance on fossil fuels and minimizing environmental impact. They also can be a double-edged sword for water and food production. While they offer cleaner energy with efficient irrigation methods, large-scale installations can disrupt farmland, and some technologies face limitations due to weather or location. Striking a balance between these factors is crucial for a sustainable future (Saleem, 2022).

There are many ways to sustainable agriculture and producing food; for example, efficient management of water, land, and agricultural resources can improve both water use and food production, or economic development can be highlighted as a key factor in achieving significant improvements in water and food productivity, but this part is focused on RECs. This factor significantly impacts those mentioned factors and significantly influences the food production index (FPI) and water productivity (WPROD). This indicated that renewable energy could improve the connection between water use, food production, and land use, contributing to a more sustainable future (Zhang et al., 2024).

RECs is a game changer for agriculture (Majeed et al., 2023). It is because of reducing costs, boosting farm productivity, and improving irrigation practices. The policies encouraging farmers to use the mentioned resources lead to mitigating climate change by reducing greenhouse gas emissions and pollution. This translates to a more stable and sustainable agricultural sector less vulnerable to energy shortages caused by extreme weather events. To maximize these benefits, future policies should actively promote REC integration into water management and food production processes across the agricultural and food sectors (Zhang et al., 2024).

5.10 THREE-PRONGED APPROACH FOR SECURING THE FOOD AND WATER FUTURE

Food and water security are critical concerns facing the planet (USDA Foreign Agricultural Service, 2023). Water is an essential source for the growth of crops and livestock, and especially in the farming and agriculture sector, water is the main factor for the reach of products and, similarly, for food. Therefore, water security and food production are related to each other. In this section, the three-pronged approach is considered for a resilient future of this theme.

Optimizing Resources in Use. Governance in the integration of land and water at the behest of the government is required to be done. This would maximize land use in farming and promote water conservation efforts. In the governance scheme, sustainable agricultural practices must be present through the acts of the government. This means practices of the scheme mandate for farmers. Further, the government should support farmers' sustainability initiatives. Coordination between farmers and government practices must be integrity that welcomes the future of security and sustained agriculture.

Concentrating on Economic Growth. Economic growth fosters sustainable practices. This factor has a significant impact on food and water productivity. One of its strategic merits is attracting financial resources to invest in sustainable farming techniques and energy-efficient technologies.

Drawing Horizon of renewable energy for agriculture. Horizon gives a strategic plan for the long term. This plan is based on green transition and designed to encourage the adoption of energy-efficient technologies, particularly those utilizing renewable sources like solar power for irrigation. In this way, all attempts to break the cycle of resource dependence turn into a renewable approach for saving future resources. This means that it makes a logical plan for the currently available resources in the future.

5.11   THE CARBON FOOTPRINT OF AGRICULTURE

The carbon footprint of agriculture refers to the total amount of greenhouse gases (GHG) emitted throughout the entire food production process, from cultivating crops to raising livestock (Jaiswal & Agrawal, 2020). This footprint significantly contributes to climate change, and understanding it is crucial for developing sustainable agricultural practices (Holka et al., 2022).

All the greenhouse gases released during farming, from growing crops to raising animals, contribute to agriculture's carbon footprint. Three main culprits are carbon dioxide, methane from animal digestion and manure, and nitrous oxide from synthetic fertilizers. However, the carbon footprint of the food supply chain and the agricultural sector leads to land-use changes and directly impacts climate change. However, understanding its relationship factors can help develop more sustainable farming methods. The statistics about this subject indicate that farm activities like raising livestock and growing crops directly generate about a fifth of global emissions. However, the impact goes beyond the farm gate (Panchasara et al., 2021).

Nitrogen-based fertilizers, deforestation for farmland, and emissions from pre- and post-production processes all contribute to the sector's carbon footprint (Gołasa et al., 2021). This under-recognized impact is due to the release of methane, nitrous oxide, and carbon dioxide, all greenhouse gases that trap heat and disrupt the planet's climate systems.

By recognizing this problem, many countries are implementing national programs to reduce agricultural emissions and achieve carbon neutrality. By taking a broader view of the food system, including processing, transportation, and consumption, it is possible to identify more solutions to mitigate climate change. This shift in perspective reveals that farm-level emissions are being overshadowed by the growing impact of supply chains and food industry practices in developed economies. Moving forward, a holistic approach considering the entire food system is crucial for effectively reducing agriculture's contribution to climate change (Thamarai et al., 2024).

5.12 RELATION BETWEEN FOOD SYSTEMS AND CLIMATE CHANGE

The food system, from farm to fork and disposal, is intricately linked to climate change. Greenhouse gas emissions contribute to food and agriculture. Almost a staggering 30% of global greenhouse gas emissions come from the entire food chain (Tubiello et al., 2022), and food production accounts for over a quarter (26%) of global greenhouse gas emissions (Poore & Nemecek, 2018). Agriculture is responsible for about half of these emissions, and it also harms the environment through deforestation, reduced biodiversity, and soil erosion. In a vicious cycle, climate change threatens food security, especially in developing countries, by worsening existing problems like hunger.

Solutions that tackle the potent greenhouse gases emitted by agriculture are needed to address this challenge. While methane emissions make up a smaller share of the total, their powerful warming effect and shorter lifespan in the atmosphere make them a prime target for immediate action. Unlike carbon dioxide, methane disappears within a decade but has a much stronger warming impact in the short term. Nitrous oxide is an even bigger threat due to its extreme potency and long life in the atmosphere. While long-term strategies for carbon dioxide and nitrous oxide are important, focusing on reducing methane emissions offers a faster way to slow climate change (Thamarai et al., 2024).

5.13 FOOD CHAIN AND SUSTAINABILITY

Food chains, where organisms transfer energy by eating each other, connect plants, herbivores, and predators. Disruptions like habitat loss can upset this balance, harming the entire web and impacting everything from human health to biodiversity. Maintaining a healthy balance in these chains is vital for a sustainable environment (Thamarai et al., 2024).

Animal-based agriculture contributes heavily to greenhouse gas emissions. Cows and sheep, for example, release methane during digestion, a potent greenhouse gas. Raising animals often requires land-use changes like deforestation, further increasing emissions. Processing meat requires significant resources like water and energy, adding to its carbon footprint (Malan et al., 2020).

In contrast, plant-based diets tend to be much more eco-friendly. Legumes like lentils and chickpeas have a significantly lower carbon footprint than animal protein sources (Nomura et al., 2023). Whole grains like wheat, rice, and fruits and vegetables generally have lower footprints. While more resource-intensive than some crops, nuts and seeds contribute less to emissions than meat. Plant-based protein alternatives like tofu and seitan offer lower footprints, too (Onwezen et al., 2021). It is important to note that specific farming practices and transportation can influence a food's overall impact, but plant-based options are generally more sustainable (Thamarai et al., 2024).

5.14 FOOD CHAIN ENVIRONMENTAL IMPLICATIONS

Centuries of clearing land for agriculture have shrunk forests and grasslands, releasing huge amounts of carbon dioxide (Yang et al., 2022). Farming emissions like methane from cows, re-growing forests, and grasslands lost could recapture massive amounts of carbon and make them a powerful tool against climate change.

Humans have been converting forests, grasslands, and other natural areas into farms for thousands of years. This has led to deforestation, with a third of the world's forests disappearing. Today, farms cover half of the earth's ice- and desert-free land. Destroying these natural habitats has significantly increased atmospheric carbon dioxide. The released CO_2, estimated at 1,400 billion tons over time, is equivalent to 40 years of current fossil fuel emissions (Thamarai et al., 2024).

Traditional methods of measuring greenhouse gas emissions often overlook "opportunity cost". This refers to the environmental benefit of using land for one purpose (farming) instead of another (forests). If they were not using this land for agriculture, it could capture some of the CO_2 that is released. Restoring these ecosystems could significantly reduce atmospheric CO_2 by allowing it to be reabsorbed by plants (Erb et al., 2018).

City parks and green spaces can also help capture carbon dioxide from the air if stakeholders try to care for them properly. Other land-use options exist, like solar or wind farms for renewable energy. However, each option has advantages and disadvantages that impact carbon storage. It needs to weigh these factors carefully before changing how humans use land.

Storing carbon in plants and soil is like taking carbon out of the atmosphere, reversing the current trend. As the stakeholders try to reduce atmospheric CO_2 levels quickly, one possible solution is to use less land for growing staple crops (like rice or wheat) (Chae et al., 2022).

Imagine if everyone on earth switched to a vegan diet by 2050. This could lead to forests and wild areas growing back, potentially capturing an additional 547 billion tons of carbon dioxide (Uddin & Kebreab, 2020). That is roughly the same amount released from burning fossil fuels in 15 years at the current rate. Researchers also believe an additional 225 billion tons could be stored in the soil, but this estimate is less certain (Zhao et al., 2018).

Another option is the "Planetary Health Diet", or EAT-Lancet diet. This plan, created by a group focused on health and sustainability, aims to benefit both people and the planet. It would not eliminate meat and dairy entirely but greatly reduce consumption. Compared to what is expected in 2050 under a "business as usual" scenario, this diet suggests eating much less red meat, poultry, and eggs, with some daily dairy consumption (Font-i-Furnols, 2023).

Following this more plant-based approach could save 332 billion tons of CO_2, about nine years' worth of the current emissions. Like the vegan scenario, researchers believe an additional 135 billion tons of CO_2 could be stored in the soil (Anwar et al., 2018).

5.15   SUMMARY

In conclusion, sustainable agriculture and food supply chains are vital for the well-being of people, the economy, and the environment. The analysis reveals that factors such as water pollution, greenhouse gas emissions, and economic challenges for farmers significantly impact the productivity and sustainability of food production systems. Addressing these issues requires the development and implementation of efficient supply chain systems that minimize environmental impacts and promote economic sustainability. By prioritizing sustainable practices, we can ensure a secure and resilient food supply for future generations, fostering a healthier planet and more robust agricultural communities. The insights and strategies discussed in this chapter underline the necessity of concerted efforts to build a sustainable food system that can adapt to and mitigate the risks inherent in modern agriculture.

REFERENCES

Aganovic, Kemal, Sergiy Smetana, Tara Grauwet, Stefan Toepfl, Alexander Mathys, Ann Van Loey, and Volker Heinz. 2017. "Pilot Scale Thermal and Alternative Pasteurization of Tomato and Watermelon Juice: An Energy Comparison and Life Cycle Assessment." *Journal of Cleaner Production* 141:514–25.

Alexandrova-Stefanova, Nevena, Kacper Nosarzewski, Zofia Krystyna Mroczek, Sarah Audouin, Patrice Djamen, Norbert Kolos, and Jieqiong Wan. 2023. "Harvesting Change: Harnessing Emerging Technologies and Innovations for Agrifood System Transformation-Global Foresight Synthesis Report."

Anwar, M. N., A. Fayyaz, N. F. Sohail, M. F. Khokhar, M. Baqar, W. D. Khan, K. Rasool, M. Rehan, and A. S. Nizami. 2018. "CO_2 Capture and Storage: A Way Forward for Sustainable Environment." *Journal of Environmental Management* 226:131–44.

Bayir, Bilgesu, Aurélie Charles, Aicha Sekhari, and Yacine Ouzrout. 2022. "Issues and Challenges in Short Food Supply Chains: A Systematic Literature Review." *Sustainability* 14(5).

Chae, Seung-Hun, Hye J. Kim, Hyeon-Woo Moon, Yoon H. Kim, and Kang-Mo Ku. 2022. "Agrivoltaic Systems Enhance Farmers' Profits through Broccoli Visual Quality and Electricity Production without Dramatic Changes in Yield, Antioxidant Capacity, and Glucosinolates." *Agronomy* 12(6).

Chandan, Anulipt, Michele John, and Vidyasagar Potdar. 2023. "Achieving UN SDGs in Food Supply Chain Using Blockchain Technology." *Sustainability* 15(3).

Cicciù, Bruno, Fernando Schramm, and Vanessa Batista Schramm. 2022. "Multi-Criteria Decision Making/Aid Methods for Assessing Agricultural Sustainability: A Literature Review." *Environmental Science & Policy* 138:85–96.

Dania, Wike Agustin Prima, Naila Maulidina Lu'ayya, and Riska Septifani. 2024. "Risk Assessment in Sustainable Food Supply Chains: A Holistic Approach by Using HOR-SWOT-ANP/AHP to Ensuring Performance Improvement." *Advances in Food Science, Sustainable Agriculture and Agroindustrial Engineering (AFSSAAE)* 7(1).

Erb, Karl-Heinz, Thomas Kastner, Christoph Plutzar, Anna Liza S. Bais, Nuno Carvalhais, Tamara Fetzel, Simone Gingrich, Helmut Haberl, Christian Lauk, Maria Niedertscheider, Julia Pongratz, Martin Thurner, and Sebastiaan Luyssaert. 2018. "Unexpectedly Large Impact of Forest Management and Grazing on Global Vegetation Biomass." *Nature* 553(7686):73–76.

Font-i-Furnols, Maria. 2023. "Meat Consumption, Sustainability and Alternatives: An Overview of Motives and Barriers." *Foods* 12(11).

Food, E. I. T. 2021. "The EIT Food Trust Report: Sustainable Food Choices and the Role of Trust in the Food Chain." EIT: Brussels, Belgium.

Ge, L., C. Brewster, J. Spek, A. Smeenk, J. Top, F. van Diepen, B. Klaase, C. Graumans, and M. de Ruyter de Wildt. 2017. *Blockchain for Agriculture and Food: Findings from the Pilot Study.* Wageningen Economic Research.

Gołasa, Piotr, Marcin Wysokiński, Wioletta Bieńkowska-Gołasa, Piotr Gradziuk, Magdalena Golonko, Barbara Gradziuk, Agnieszka Siedlecka, and Arkadiusz Gromada. 2021. "Sources of Greenhouse Gas Emissions in Agriculture, with Particular Emphasis on Emissions from Energy Used." *Energies* 14(13).

Gonçalves, Amélie, and Thomas Zeroual. 2017. "Logistic Issues and Impacts of Short Food Supply Chains: Case Studies in Nord – Pas de Calais, France." *Toward Sustainable Relations Between Agriculture and the City.* pp. 33–49 in, edited by C.-T. Soulard, C. Perrin, and E. Valette. Cham: Springer International Publishing.

Holka, Małgorzata, Jolanta Kowalska, and Magdalena Jakubowska. 2022. "Reducing Carbon Footprint of Agriculture – Can Organic Farming Help to Mitigate Climate Change?" *Agriculture* 12(9).

Irani, Zahir, and Amir M. Sharif. 2016. "Sustainable Food Security Futures." *Journal of Enterprise Information Management* 29(2):171–78.

Jaiswal, Bhavna, and Madhoolika Agrawal. 2020. "Carbon Footprints of Agriculture Sector." *Carbon Footprints: Case Studies from the Building, Household, and Agricultural Sectors.* pp. 81–99 in, edited by S. S. Muthu. Singapore: Springer Singapore.

Kamble, Sachin S., Angappa Gunasekaran, and Shradha A. Gawankar. 2020. "Achieving Sustainable Performance in a Data-Driven Agriculture Supply Chain: A Review for Research and Applications." *International Journal of Production Economics* 219:179–94.

Krishnan, Ramesh, Renu Agarwal, Christopher Bajada, and K. Arshinder. 2020. "Redesigning a Food Supply Chain for Environmental Sustainability – an Analysis of Resource Use and Recovery." *Journal of Cleaner Production* 242:118374.

Krithika, L. B. 2022. "Survey on the Applications of Blockchain in Agriculture." *Agriculture* 12(9):1333.

Kumar, Anish, Sachin Kumar Mangla, and Pradeep Kumar. 2022. "An Integrated Literature Review on Sustainable Food Supply Chains: Exploring Research Themes and Future Directions." *Science of the Total Environment* 821:153411.

Kumar, Anish, Sachin Kumar Mangla, and Pradeep Kumar. 2024. "Barriers for Adoption of Industry 4.0 in Sustainable Food Supply Chain: A Circular Economy Perspective." *International Journal of Productivity and Performance Management* 73(2):385–411.

Lee, Hyesoo, Sehun Choi, Euichan Kim, Ye-Na Kim, Jihyun Lee, and Dong-Un Lee. 2021. "Effects of Pulsed Electric Field and Thermal Treatments on Microbial Reduction, Volatile Composition, and Sensory Properties of Orange Juice, and Their Characterization by a Principal Component Analysis." *Applied Sciences* 11(1).

Leng, Kaijun, Ya Bi, Linbo Jing, Han-Chi Fu, and Inneke Van Nieuwenhuyse. 2018. "RETRACTED: Research on Agricultural Supply Chain System with Double Chain Architecture Based on Blockchain Technology." *Future Generation Computer Systems* 86:641–49.

Majeed, Yaqoob, Muhammad Usman Khan, Muhammad Waseem, Umair Zahid, Faisal Mahmood, Faizan Majeed, Muhammad Sultan, and Ali Raza. 2023. "Renewable Energy as an Alternative Source for Energy Management in Agriculture." *Energy Reports* 10:344–59.

Malak-Rawlikowska, Agata, Edward Majewski, Adam Wąs, Svein O. Borgen, Peter Csillag, Michele Donati, Richard Freeman, Viet Hoàng, Jean-Loup Lecoeur, Maria C. Mancini, An Nguyen, Monia Saïdi, Barbara Tocco, Áron Török, Mario Veneziani, Gunnar Vittersø, and Pierre Wavresky. 2019. "Measuring the Economic, Environmental, and Social Sustainability of Short Food Supply Chains." *Sustainability* 11(15).

Malan, Hannah, Ghislaine Amsler Challamel, Dara Silverstein, Charlie Hoffs, Edward Spang, Sara A. Pace, Benji L. Malagueño, Christopher D. Gardner, May C. Wang, Wendelin Slusser, and Jennifer A. Jay. 2020. "Impact of a Scalable, Multi-Campus 'Foodprint' Seminar on College Students' Dietary Intake and Dietary Carbon Footprint." *Nutrients* 12(9).

Mancini, Maria C., Davide Menozzi, Michele Donati, Beatrice Biasini, Mario Veneziani, and Filippo Arfini. 2019. "Producers' and Consumers' Perception of the Sustainability of Short Food Supply Chains: The Case of Parmigiano Reggiano PDO." *Sustainability* 11(3).

Moller, Björn, Lorenzo Giacomella, Anna Kirstgen, Kerstin Pasch, Kemal Aganovic, Ewa Dönitz, and Ariane Voglhuber-Slavinsky. 2023. Local food systems. Recipes for future proof business models. Fraunhofer ISI. Case studies on how innovative food processing technologies can be used economically to boost local food value chains.

Nomura, Marika, Miwa Yamaguchi, Yuji Inada, and Nobuo Nishi. 2023. "Current Dietary Intake of the Japanese Population in Reference to the Planetary Health Diet-Preliminary Assessment." *Frontiers in Nutrition* 10:1116105.

Onwezen, M. C., E. P. Bouwman, M. J. Reinders, and H. Dagevos. 2021. "A Systematic Review on Consumer Acceptance of Alternative Proteins: Pulses, Algae, Insects, Plant-Based Meat Alternatives, and Cultured Meat." *Appetite* 159:105058.

Ozkan-Ozen, Yesim Deniz, Yigit Kazancoglu, and Sachin Kumar Mangla. 2020. "Synchronized Barriers for Circular Supply Chains in Industry 3.5/Industry 4.0 Transition for Sustainable Resource Management." *Resources, Conservation and Recycling* 161:104986.

Panchasara, Heena, Nahidul H. Samrat, and Nahina Islam. 2021. "Greenhouse Gas Emissions Trends and Mitigation Measures in Australian Agriculture Sector – a Review." *Agriculture* 11(2).

Poore, Joseph, and Thomas Nemecek. 2018. "Reducing Food's Environmental Impacts through Producers and Consumers." *Science* 360(6392):987–92.

Rehman, Abdul, Muhammad Farooq, Dong-Jin Lee, and Kadambot H. M. Siddique. 2022. "Sustainable Agricultural Practices for Food Security and Ecosystem Services." *Environmental Science and Pollution Research* 29(56):84076–95.

Reina-Usuga, Liliana, Tomás de Haro-Giménez, and Carlos Parra-López. 2020. "Food Governance in Territorial Short Food Supply Chains: Different Narratives and Strategies from Colombia and Spain." *Journal of Rural Studies* 75:237–47.

Rejeb, Abderahman, John G. Keogh, and Karim Rejeb. 2022. "Big Data in the Food Supply Chain: A Literature Review." *Journal of Data, Information and Management* 4(1):33–47.

Saleem, Muhammad. 2022. "Possibility of Utilizing Agriculture Biomass as a Renewable and Sustainable Future Energy Source." *Heliyon* 8(2).

Sharma, Janpriy, Mohit Tyagi, and Arvind Bhardwaj. 2023. "Valuation of Inter-Boundary Inefficiencies Accounting IoT Based Monitoring System in Processed Food Supply Chain." *International Journal of System Assurance Engineering and Management* 15(4):1374–1396.

Silva, Beatriz Q., Eva Kancirova, Milena Zdravkovic, Uday Batta, János-István Petrusán, Kerstin Pasch, Kemal Aganovic, Marta W. Vasconcelos, and Sergiy Smetana. 2024. "Sustainable Food Chains Designed for Optimised Resource Use: Optimising Downscaled Food Chains for Sustainable Resource Use: A Comprehensive Case Study on Tomato Juice." *Journal of Cleaner Production* 450:141879.

Subramanian, Nachiappan, Atanu Chaudhuri, and Yaşanur Kayikci. 2020. "Blockchain Applications in Food Supply Chain." *Blockchain and Supply Chain Logistics: Evolutionary Case Studies.* pp. 21–29 in, edited by N. Subramanian, A. Chaudhuri, and Y. Kayıkcı. Cham: Springer International Publishing.

Tao, Qi, Hongwei Ding, Huixia Wang, and Xiaohui Cui. 2021. "Application Research: Big Data in Food Industry." *Foods* 10(9).

Thamarai, P., V. C. Deivayanai, A. Saravanan, A. S. Vickram, and P. R. Yaashikaa. 2024. "Carbon Mitigation in Agriculture: Pioneering Technologies for a Sustainable Food System." *Trends in Food Science & Technology* 147:104477.

Tubiello, Francesco N., Kevin Karl, Alessandro Flammini, Johannes Gütschow, Griffiths Obli-Laryea, Giulia Conchedda, Xueyao Pan, Sally Yue Qi, Hörn Halldórudóttir Heiðarsdóttir, Nathan Wanner, and others. 2022. "Pre- and Post-Production Processes Increasingly Dominate Greenhouse Gas Emissions from Agri-Food Systems." *Earth System Science Data* 14(4).

Uddin, Md E., and Ermias Kebreab. 2020. "Impact of Food and Climate Change on Pastoral Industries." *Frontiers in Sustainable Food Systems* 4:543403.

USDA Foreign Agricultural Service. 2023. *Field Guide: A Three-Tiered Approach to Increasing Sustainable Water and Food Security in the APEC Region.* USDA.

Vittersø, Gunnar, Hanne Torjusen, Kirsi Laitala, Barbara Tocco, Beatrice Biasini, Peter Csillag, Matthieu D. de Labarre, Jean-Loup Lecoeur, Agnieszka Maj, Edward Majewski, Agata Malak-Rawlikowska, Davide Menozzi, Áron Török, and Pierre Wavresky. 2019. "Short Food Supply Chains and Their Contributions to Sustainability: Participants' Views and Perceptions from 12 European Cases." *Sustainability* 11(17).

Yang, Dan, Zhenyue Liu, Pengyan Zhang, Zhuo Chen, Yinghui Chang, Qianxu Wang, Xinyue Zhang, Rong Lu, Mengfan Li, Guangrui Xing, and Guanghui Li. 2022. "Understanding Relationships between Cultivated Land Pressure and Economic Development Level across Spatiotemporal Characteristics: Implications for Supporting Land-Use Management Decisions." *International Journal of Environmental Research and Public Health* 19(23).

Zhang, Xiaoli, Xinling Wang, Dingwen Si, Hong Zhang, Mohammed Moosa Ageli, and Grzegorz Mentel. 2024. "Natural Resources, Food, Energy and Water: Structural Shocks, Food Production and Clean Energy for USA in the View of COP27." *Land Degradation & Development* 35(7):2602–13.

Zhao, Huili, Huijie Zhang, Abdul Ghaffar Shar, Jifei Liu, Yanlong Chen, Songjie Chu, and Xiaohong Tian. 2018. "Enhancing Organic and Inorganic Carbon Sequestration in Calcareous Soil by the Combination of Wheat Straw and Wood Ash and/or Lime." *PLoS One* 13(10):e0205361.

6 Blockchain Applications in Sustainable Agriculture

Sara ravan Ramzani, Ahmad Abu-Alkheil,
Vahab Esfandani, and Reza Ahmadi

6.1 INTRODUCTION

Being decentralized and immutable, meaning that no single entity may change or control it, are the main characteristics of blockchain technology. The recognition that it now enjoys underlines its potential to transform businesses globally.

Divisional digital records where data spread over various computer areas of business activity or nodes are the model of operation for blockchain. Through formalizing the process within each unit and cross-referencing the recorded business transactions with previous postings, block technology sets up these transactions into multiple layers of blocks in a sequential and non-disruptive order.

According to Priyadarshini (2019), the divisional system employed to run it, makes sure that no office or unit has complete authority and control over the operations inside the functional areas. This promotes openness and expansion of operations.

Technology provides answers to pressing problems in the agriculture and food processing industries including inadequate supply chain logistics, food pilferage, and failures in transparency. Given that blockchain ensures privacy, security, and an open system over the materials of the activities in the ledger, it has the power to endure the demands of the food sector. The consequence of this function can lead to a change in the position of food processes in relation to production, distribution, and consumption, consequently ending up in a reliable sustainable and seamless system.

The detrimental effects of blockchain technology on data storage and transaction management as well as its non-centralized and limited security measures against unauthorized individuals have been extensively documented. Some of the major areas where this technology is used are banks, healthcare services, and supply chain functional aspects by delivering effective role to improve openness, credibility, trust, and reliability through all the functioning areas of agriculture in relation to food supply chain.

Due to Ge et al. (2017), the usage of the cryptographic features embedded into the blockchain has improved the seamless monitoring and recording of significant data along with other areas related to invaluable seed items, crop development, product distribution, and sales.

DOI: 10.1201/9781003609865-6

6.2 INTELLIGENT FARMING

New methods for agricultural applications were produced through the integration of blockchain systems, Internet of Things (IoT) components, and other related technologies. It creates a way for safe as well as decentralized forums for data management and acquisition.

The use of this tool has made farmers to be more intimate with and improve in their approach to the agricultural practices, mostly in watering, application of food supplement, and management of pests with positive impact on their productivity and remarkable effectiveness in the use of resources (Sendros et al., 2022).

6.2.1 Food Supply Chain

As previously stated, openness and validity of food supply logistics are some of the major benefits resulting from the use of block technology which permits all authorized persons to be able to supervise and vouch activities within the agricultural areas from the production areas to the end user. The accuracy of information produced by blockchain made food fraud to be minimized and reduction in poisonous food. Hence, this enhances trust between producers and the end user by securing and preservation of transactional data on immutable record platforms (Van-Wassenaer et al., 2021).

6.2.2 Insurance for Agriculture

Insurance systems in agricultural sector are more effective with clarity using blockchain technology most especially where models adopted are based on indexing. Adopting this technology could make the processes to be configured with automation and resulting to increasing efficiency which could add more to the output of the farmers with high turnover on the assumption that all things being equal there would be favorable atmospheric weather coupled with encouraging farm produce. When this happens, workload will drastically reduce and the capability to face issues around climate would be bearable (Sendros et al., 2022).

6.2.3 Agricultural Product Transactions

With the use of this technology farm, product transactions can be kept more secure and transparent and associates can continue to do business with one another without the need for middlemen. Effective arrangements introduced on blockchain activities with uninterrupted and enforceable mutual understanding subsisting among the producers and consumers ensure trust and reliability in farm produce industry (Sendros et al., 2022).

6.2.4 Smart Agriculture

Innovative scientific methods for agricultural activities are made possible by the networks together with other software to run data analytics and the IoT (Figure 6.1). It is

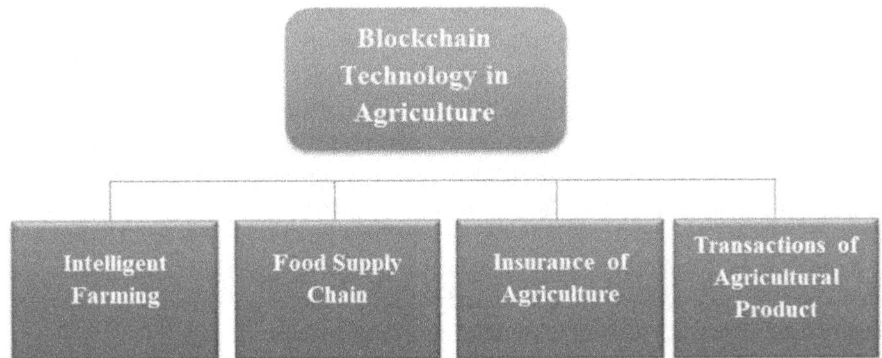

FIGURE 6.1 Blockchain technology in agriculture sector.

made possible by providing prompt and careful information on weather conditions, agricultural output quality, and the effectiveness of material use. As stated by Van-Wassenaer et al. (2021), farm produce operators might take the advantage of using blockchain technology for purposes like information processing, administration, and distribution. This enables them to have an optimal usage of a wide range of information sources assisting them in decision-making processes, ultimately leading to higher performance, lesser unproductive goods, and overall sustainability throughout the farm produce activities. The widespread deployment of blockchain system and its high robust potential to replace the conventional approach with positive and beneficial impacts on other difficulties have lowered constraints faced by the agricultural sectors and farm produce areas.

By guaranteeing openness, audit trailing, and activeness, this tool encourages the creation of sustainable, flawless, and solid food processes. In an ordinate number of prominent industry players including Alibaba, JD.com and Walmart were among the first to implement blockchain technology auditing for projects, hence highlighting its acceptability and broad use in the agricultural produce segment.

Adoption of farm produce and food processes using integrated blockchain technology has an improvement which encourages farmers to actively use materials, through supervision on the health of the farm produce and implementing specific farm processes. In addition, this technology helps the clearing and reliability of administering farm produce logistics, thereby bringing down reducing the likely barriers in relation to food poison, contaminated and adulterated food items.

The introduction of agricultural insurance program on blockchain technology supports the farmers to mitigate risk elements. These programs make use of real-time processed information and smart arrangements to facilitate the filing process and establish capability to disallow any unfavorable weather disturbances (Van-Wassenaer et al., 2021).

The adoption of blockchain technology in the farm produce sector and food processes has led to have upper hand in relation to the reason for sustainability, transparency, and ethical practices in the production of food items. The application of this technology in farm produce has more widespread in providing visibility and traceability in transactions. The unstructured system provides room for uninterrupted internal dealings among colleagues and bringing down the cost.

Safekeeping of records and dependable processing systems are guaranteed with the use of this technology in all the units where farm produce activities are being taking place from the primary stage to the end user, such that any stakeholder has opportunity to access adequate information toward adopting strategic maximization of the wealth standing within the segment of the market position (Kamilaris et al., 2019).

Even though this technology has a general acceptance, it is still facing some barriers which are gaps calling for a new study to reveal the importance of various types of agricultural practices along with their cost implications.

The current mode of operation in place has been carried out without complications ensuring flexibility and minimizing disruptions. The cost involvement in the processing of information and withdrawal poses major challenges more noticeably in small farm produce sectors (Antonucci et al., 2019). Additionally, the capability in dealing with information processing is very relevant when information is mounting up in volumes, which involves many parties in different locations.

Globalization and integration with other sectors are very necessary to hike up the usage of blockchain technology to bring out the hidden strength and potentials to foster sustainability in the ecosystems, openness, and generally accepted farm produce practices. The essence of this study is to bring to limelight the functions of blockchain in sustainability of farm produce with reference to improving trust, openness, and establishing audit trails in supply logistics. A close study of importance, barriers, and challenges revealed that certain inherent element of blockchain when trying to apply it is in solving operational issues and uplifting more economical ways of using materials and efficient approaches on farm produce.

To round up this discussion, the use of non-renting or privately owned technology can allow more opportunity to business re-engineering, transparency, trust, mutually profitable dealings, and Corporate Social Responsibility in relationship to farm produce processing and supply logistics (Demestichas et al., 2020).

6.3 OBJECTIVES

This study was packaged to reveal the outcome of the examination of the adoption of blockchain technology in farm produce and food practices. In reporting this study, the introduction aspect was cross related to other sections of the package. Following the report, the theoretical philosophies of this tool and the potentials in the farm produce sector were discussed.

The scope of this research was from these countries, that is, Germany, Pakistan, and Hungary, revealing life demonstrations of how the use of blockchain technology in the farm produce industry and its effects on farm produce practices. Variables such as distribution logistics, information processing, and confirmatory processes were put into consideration under each case that was examined in this research.

Furthermore, the study was contrasted with other practices offering solutions just like blockchain technology, highlighting its benefits and challenges when applying it from the basics.

The end point of this research is an integration of ideas emanated from the study giving directions toward any intending future studies, narrowing down to specifics on how blockchain might re-engineer farm produce sector and farm produce practices.

6.4 BLOCKCHAIN AND USE IN AGRICULTURE INDUSTRY

The aim of this section is to examine the applicability of blockchain technology (BCT) in the farm produce sector from the point of view of diverse research and the analyzing of its adaptation to establish reliability, workability, and audit trail throughout all the logistical operational areas. This aspect examines deeply the applications of the "said tool" as stated focusing on the "said sector" also, by expatiating on the factors underlining development as previously discussed.

Furthermore, it reiterated the impact of the breaking of ground by blockchain on the potentiality of scientific farm produce by discussing the approach of resolving key areas of concern, that is, openness, auditing, and farm produce protection.

One of the key aspects of blockchain is called "block" for the role of building up of chains by recording the hash code in the system of first in to be recorded at first basis. There is an agreed geometrical computation of the hash code in each block of the blockchain which includes a header consisting of detailed information about the organization ensuing this idea.

It is to be noticed that the block carries extra information and taking up of blockchain technology resulted in a remarkable surge as a consequence of unplanned digital currency called "Bitcoin" as named by Satoshi Nakamoto in 2008 (Motta et al., 2020).

The root of Bitcoin is traceable to blockchain technology (BCT) due to its features and the platform into which it is founded. The operations of this tool are unstructured, and recording approach is divisional whereby main transactions are the nodes linking other subsidiary-associated postings individually.

Distributed ledger technology (DLT) that is the technological infrastructure and protocols which allow simultaneous access, validation, and updating of records across all networks of database ensuring processed information channeled across certain operational approaches (Duan et al., 2020). The spread approach adopted by the ledger recoding led to the exposure of information not being protected or secured, thereby made it difficult to have an approved recognized appointed institution to monitor and be responsible as ownership of data.

The distributed approach employed in blockchain made it difficult to establish reliability within the nodes (Kouhizadeh & Sarkis, 2018). The two elements namely non-centralized and non-concentrated ledger are key measures to information processing system and efforts to modify processing information. Mis-posting of information or unapproved update would lead to interruptions in connections (Varriale et al., 2020).

Non-centralized peer-to-peer (P2P) approach ensures recovery of any missing data; hence, blockchain which is a non-centralized digital system is meant for posting of data that uses cryptographic approach to ensure security and validity of transactions (Vangala et al., 2020).

Satoshi Nakamoto was the pioneer in taking advantage of blockchain technology (BCT) in the Bitcoin application. Yet Bitcoin opened the gate for alternative digital currencies and ensured flexibility under a range of conditions (Rocha et al., 2021). Specifically in relation to smart farm produce, BCT is capable to update transparency and audit and provides impressive results in distribution logistics (Sharma et al., 2021). Due to its natural production of farm produce and other raw materials, agricultural produce plays a significant role in the global economy.

From the point of view of increasing worldwide population, it becomes very necessary to adopt a set of strategies to encourage increasing farm produce activities and achieving planned result (Song et al., 2021).

The farm produce sector is oblivious of the BCT as a demonstration of practical and speedy solutions for reliability, productivity, and auditable avenues; such that they have benefited remarkable things as a result of current high-level technological innovations (Sharma et al., 2020).

This technology (BCT) is based on data electronically with adequate protection without volatility and its importance can be attributed to the capability of monitoring and verification of ledger postings, the dogmatism of imputed processed information, and openness, coupled with high protection (Demestichas et al., 2020). Sincerity and openness form a key function in the farm produce industry, most especially in the aspect of relentless and reliability of farm produce logistics (Liu et al., 2021). Adoption of this technology into farm produce operations creates opportunity for improvement of reliance and openness in all farm produce logistics areas.

The integration of this tool has influence on performance, supervision, productivity, and audit trail of produce to the extent of economically managing the uncertainty related to contaminated food substances that can lead to ill-health (Kramer et al., 2021). Introducing a monitoring and detective approach that could specifically control with assurance and protection of farm produce is very necessary which has a remarkable positive effect on end users, constituted organizations, and other interest groups within the industry.

The group of researchers (Kamble et al., 2020) stood on a stand that BCT offers more effective and efficient approaches to make all things happen. It has been established that "Smart agriculture" relates to an agricultural system that adopts up-to-date technological approaches to increase productivity of crops, minimizing wastages and environmental sustainability enhancement (Marzougui et al., 2023). By introducing artificial intelligent sensors, unmanned aerial vehicle (UAV) and various technological software to seek reliable processed information on the circumstance of the soil in its natural being, atmospheric situations, and the behavior of farm produce.

The material has motivating influence on the process at which the decision is being made in relation to planting of produce, how they are being fertilized and harvesting (Dey & Shekhawat, 2021). The administration of biological produce being monitored by technology called "BCT" in environmentally conscious farm produce generates encouraging output.

To employ this scientific approach can support openness and increase in performance of distribution logistics, such that the risk of uncertainty on pilferage and ensuring that end users are supplied with adequate information in relation to avenues where farm produce are being sorted (Mukherjee et al., 2022). In addition, certain definite farm produce can adopt the technology to support the process of making decision based on the timing optimization, methodology, use of fertilization, and harvesting (Elijah et al., 2018).

At the starting point, blockchain technology is capable to support smart agriculture (Antonucci et al., 2019), considering the benefits of close supervision in relative terms. It is therefore necessary to focus on the validity and efficiency of blockchain technology (BCT) in the farm produce industry in relation to honesty, performance, and audit trail in all over farm produce distribution logistics.

6.5 USE OF BLOCKCHAIN IN SOUTH AMERICA FOR SUSTAINABLE AGRICULTURE

The last few aspects of this report were concentrated on blockchain technology and its importance to farm produce industry giving attention to its impact on openness, audit trail, and encouraging performance throughout the distribution logistics.

The aspect of this report would focus on the use of the software in South America enhancing agricultural sustainability on the current platform. Attention shall be toward the challenges and opportunities to consider on various farm produce approaches by considering the essence of security and at the same time ensuring positive productivity of the atmospheric conditions to be for an unforeseeable period in the future, lifestyle of inhabitants, coupled with the economic activities.

The advantages and disadvantages are also being considered being important used in the country for farm produce protection for sustainability (Westerkamp et al., 2020; Sharma et al., 2022). However, there are barriers to achieve the present and future expectations of the environment and at the same time ensuring prospects, sustainable atmosphere, promising favorable and fair lifestyle, and active business activities (Mba et al., 2020).

Farm product's sustainability performs remarkably effective in addressing the four key features including farm produce protection, pathways for usage along with reliability. For confirmation, the unit also observed these elements as explained by the UN's Food and Agriculture Organization (FAO) as part of their economic activities with Forestry and Evaluation Working Report with effect from the year 1962. Presently, we experienced re-engineering in farm produce sector and distributing logistics which is a combination of many sub-sections taking place in all aspects from the start to finishing with final user (Astill et al., 2019). Involving many productive operations which include farm produce, collection, manufacturing, protection, distribution couple with selling, and the intention is to affirm the efficiency of the protection put in place for the farm produce with insecurity (Reyes et al., 2020).

Agricultural sustainability promotion is connected to the following robust technologies which have data storage capacities. According to Ordóñez et al. (2023), three key approaches are nowadays being used to record and publicize agricultural produce: blockchain, artificial intelligence (AI), and machine learning (ML). In 2009, Satoshi Nakamoto brought to the limelight the blockchain with storage and interchanging functions within the nodes of a distributed network, thereby replacing a constituted body (Treiblmaier, 2019).

These nodes are part of posting instituting from recording of transactions and well known for. It is notable for being a technological self-dependent without the presence of physical person and unalterable which is being recognized within the distributed database and hives room for interchanging and access to processed information (Zhu & Kouhizadeh, 2019).

This software is being adopted in circumstances where privacy, reality, and laid-down policies are being observed (Reyes et al., 2020) ensuring unchangeableness, openness, and protection through eliminating in-between agencies and data processing privacy. The equipment has a significant effect on audit trail coupled with distribution logistics on farm produce (Li et al., 2021; Song & Li, 2021).

6.6 USE OF BLOCKCHAIN IN GERMAN AGRICULTURE

The previous aspects of this report were concentrated on blockchain technology in South American and its importance to farm produce industry giving attention to its impact on openness, audit trail, and encouraging performance throughout the distribution logistics and the benefits, challenges being experienced from the use of blockchain in its farm produce activities.

Following up on the last report, we will examine how this tool is used in Germany's agricultural produce in relation to legal, manufacturing, and cultural issues that affect its use in a country where a lot of emphasis has been placed on connecting activities taking place in distribution areas. The essence of this unit is to investigate the level of performance coupled with the fitness of the mechanism of the tool and carry out an appraisal of the post effect of economic activities and to resolve observations relating to sustainability peculiar to Germany. Part of the studies carried out in Germany have been additions which had offered gainful materials in relation to this mechanism and the usage of it coupled with the mechanism connecting vertical distribution logistics especially (Kramer et al., 2021).

From the point of view of deep knowledge on this research in relation to legal status, business setting and elements pertaining to social issues made room for application of the use of block technology in Germany under the farm produce and sectors. Data analysis of this country revealed that the study has a positive impact on the knowledge, improvement, and fitness of various aspects of the blockchain technology platforms (BCTPTs) in monitoring the distribution logistics and developing strategic alliances.

In addition, legal, corporations, and researchers are privileged to important information looking from the point of Germany under the assumption of likely opportunities and challenges adopting the mechanism in farm produce sector and distribution logistics (Groombridge, 2020), the consequence of which would highlight the essence of reality as shown in the report of the research.

The features of farm produce logistics are noted by divisional units where a constituted organization has been charged with the responsibility of leading and making provision on its branding. The organization responsible for supply chain takes charge of interconnecting operations (De Vries, 2020). This mechanism is being seen to be developing in activities irrespective of divisional outlets most especially in supervision, audit trailing, farm produce, and making available materials in relation to source (Hanf & Dautzenberg, 2006). Different sectors with farm produce areas, financial, health care, manufacturing, coupled with transportation, have already adopted blockchain technology (BCT) with associated accessories.

The divisional foundations of this equipment are prone to lead to re-engineering in supply chain management (SCM) (Behnke & Janssen, 2020). Nevertheless, we have not carried out deep studies for the administration of farm produce supply chain networks (SCN); however, the effect of blockchain technology (BCT) on the collaboration of distribution logistics becomes open due shortage of studies in relation to the impacts on BCTs for collaboration and coordination.

Furthermore, many studies were not available on how behavior change techniques (BCTs) which could reduce challenges of how decision-making is being rationally taken, likely attitude and inequality to get to data, that would have made the distribution logistics to perform in an improved manner.

Putting in perspective a proposal to introduce blockchain technology (BCT), it becomes necessary to critically have accessibility to the gainful aspects of the distribution logistics and operations and at the same time recognize the total cost of ownership (TCO).

This TCO is the combination of the initial capital expenditures (CAPEX) and ongoing operating expenditures (OPEX). This study focuses on appraisal of the economical impact of blockchain technology (BCT) in farm produce, produce activities, and distribution logistics concentrating on the report on appraisal of the specific usage of the tools.

In addition, consideration was given to individual environmental blockchain technology sustainability for this area (Deloitte, 2020). If economic issues are to be analyzed under the distribution logistics with efficiency, it is necessary to consider current developments in relation to geometrical movement of rate of consuming energy on BCT, according to Treiblmaier (2018). This increase can be attributed to the need for Bitcoin mining operation all over the blockchain system, which was because of unperforming proof-of-work (PoW) agreement system. Jović et al. (2020) state that currently, any aspect of the dealings on the Bitcoin blockchain makes use of 668 kWh of energy consumption which is a replica of family domestic use in the US within 23 days. Presently, various debates are on blockchain technology sustainability, especially because of proof-of-work (PoW) process (Huh & Kim, 2019; Vranken, 2017).

In case a general unrealistic algorithm is attached to blockchain-based consensus protocol technology (BCTPT), resultant effect could lead to operational expenses (OPEX). Other studies dug into effects of various likely sustainability blockchain agreed protocols, which are susceptible to take the position of proof of work (POW), in conjunction with related economic consequences (Davidson et al., 2018).

6.7 USE OF BLOCKCHAIN IN PAKISTANI AGRICULTURE

This aspect discussed the barriers and opportunities of adopting blockchain technology in the farm produce industry of Pakistan aiming at assessing flexibility, provision of workable ideas and giving guidelines for the next generation to exploit more on the benefits of blockchain technology in Pakistani's farm produce industry. Sialkot, Pakistan, well-noted person for the contributions in farm produce industry, whose contributions have tremendously by record yielded a yearly export to the tune of $900 million, occupying a line space of 642,624 acres. However, traditional farm produce activities, which resulted in environmental downwardness on healthiness from the point of wastages on consumption suppressed this aspect which this research is bent to discuss in order to understand the essence of traditional items (Khan et al., 2022).

This issue is to promote the use of blockchain technology to control and document farm produce activities starting from the production to the end user (Farooq et al., 2024). Also, the research looks at the consequences of farm produce pesticides coupled with ingredients on seed outcomes at the same time encouraging sustainability by adding of Internet of Things (IoT) sensory systems and the use of low-energy adaptable clustering hierarchy (LEACH) system to maximize the energy usage.

Initial studies revealed that there were remarkable differences arising from soil's natural features due to scientific approaches especially on temperature and smoke levels (Li et al., 2021). The objective of this research is to develop farm produce protection and make the best use of the uncertainties by putting into consideration end user's options and making sure that proper recording is observed on the distribution logistics channels. Without struggle, we applied individual blockchain technology (Khan et al., 2022). The inclusion has the capability to improve farm produce activities and soliciting for global recognition on health care and farm produce processes.

Agriculture practitioners approve blockchain technology in farm produce sectors as an accounting recording of activities and important data usage within them and their agencies at each unit (Bermeo-Almeida et al., 2018). Chofreh et al. (2019) argued that this forum supports recording of seed administration and faces challenges of global health care coupled with issues around farm produces.

Dujak and Sajter (2018) supported IoT devices in application of blockchain equipment on farm produce activities, crop varieties/planting, the chemical measures to be used, and the gathering of the crop's yields.

From various researches in relation to developing countries, old approach has not been eradicated in their farm produce practices as manually is more of being in use at each stage of operation (Galvez, 2018). Contaminated food is a key barrier to guarantee food items to be classified as being high quality (Gurtu & Johny, 2019). This nullifies the objectives of sustainability in farm produce sector to access important data which also impaired all stakeholders of any type (Ullah, 2021).

Currently, activities require blockchain-powered resolutions to support the performance of farm produce activities which accounted for the 2009 discussion on smart agriculture as a workable approach to revamp farm produce and support issues in the environment (Awan et al., 2021).

Blockchain is a technology that emerged with the strength of improving farm produce logistics, capable of eradicating destructions, ensuring food protection, and enhancement of openness. The IoT is a more advanced environmental conducive approach to control destructive insects, water gauge, and other ecological elements by thinking ahead of any form of natural deprivation (Awan et al., 2020a). The adoption of the IoT in farm produce activities is being efficient approach in which its combination with blockchain technology is to protect the agricultural operations industry (Alkahtani et al., 2021) and improvement on the farm produce productivity.

Farm produce and soil control measures give opportunity to a more advanced approach to applicability within the confines of farm produce (Khalil & Ahmed, 2024), and the consequential effect of this leads to enhancement of farm producers' productivity and efficiency (Awan et al., 2020b). This device collects information virtually and communicates it to a structured environment to transparent on electronical or graphical software (Alkahtani et al., 2021).

Most software linked to internet, upon connecting them to routers would constantly transmit messages from where they are based to the atmosphere. However, the use of smart device observed an issue about protection and safety (Awan et al., 2019).

Instituting a constituted authority of trust that monitors data transmission into the atmosphere like data warehouse would reduce a lot of the concerns (Ehsan et al., 2022). Each aspect of a blockchain has audit trail of every transaction which would show historical, sequent, and full characteristics of data in relation to each unit. When information on trading moves from the node, it is apparent for anyone to see with connectivity and its evidence is noticed by the receiver which is verifiable when exchanging the information (Farooq et al., 2023).

Throughout the verification broadcast, the operational node is connected to other active notes in the system, making the encoder to have the expected response from the decoder which is readily available. The reliability of the transaction is then confirmed by Raj, Kumar (Ullah, 2021). According to Hakim et al. (2024), a remarkable feature of blockchain system is the capability to guarantee suppliers' development protection over farm produce records and the sharing of information with relevant strategic partners.

6.8 USE OF BLOCKCHAIN IN HUNGARY'S AGRICULTURE

This chapter shall progress on the barriers and benefits associated with the application of blockchain technology in farm produce sector of Hungary which aims to introduce strategies that would reduce the barriers, introduce remedies, and highlight proposed studies on total measures in which this mechanism can fully be mobilized to tackle farm produce activities and food distribution logistics in Hungary.

Alobid et al. (2022) carried out a study in Hungary in the area of blockchain technology on farm produce sector with the aim of stretching the future advantages on farm business activities, farm produce logistics and control. The study was able to appraise a sample of 79 papers by looking into the present situation, related barriers, and likely impact of blockchain-based technology on farm produce. Farmers Edge is a typical example of agribusiness organization engaging this mechanism to support farm produce activities globally.

More people drew importance to Farmers Edge's digital farm produce remedies for their state-of-the-art artificial intelligence technology. The research stretched the importance of blockchain mechanism to promote openness and productivity in farm produce industry by grouping processes for suppliers and end users. Some consumers took a careful look at the food items that they consume, the problems in the distribution and having quality at stake. The use of this mechanism enhances easy accessibility of processed information, preventing consumers from surreptitiously modifying or hiding it (Galvez et al., 2018).

Adopting this system, the buyer could confirm the animals' well-being, carriage details, period of storage, and other identities by allocating a QR code symbol to each of the beef packages (Vaia, 2020; Ganne, 2018). Without agreement from all connected persons, it is not possible for anyone to rectify the processed stored information on the blockchain (Zhang et al., 2019). Using this strategy gave room to protection on adopters against unscrupulous persons and protection on issues like fraud and ensured reliability. To put trust in a system, there must be assurance of integrity and access to authorized users (Casino et al., 2019). Adjusting prices would demand prices, how elements of costs are being arrived at through a recognized team of cloud developers with avoidable rates and a list of farm produce investments; for example, an organization must have the capability to provide quality items which "is both well-crafted and meets the required data records" (Blanchard, 2021). Maintaining an ongoing concern while also developing a relationship may present a challenge that lasts for approximately two years.

"Initial coin offers (ICOs) are a type of crowdfunding that utilizes cryptocurrency as opposed to conventional payments to facilitate the connection between small business owners and investors, hence aiding in the expansion of their firms" (Ante et al., 2018). The approach of introducing new ideas into productive areas includes efforts to talk to persons to engage in a gainful trading. For the fact that initial coin offerings are not being controlled by any arm of government, dealers could ignore any unrecognized legal cost aside from the solicitor's engagement costs (Chohan, 2019).

Smart contracts, developed by software engineers and linked to the network, remarkably simplify the food delivery and payment collection processes. The immediate execution of operations upon product readiness leads farmers not to endure weeks or months of waiting for their money to be repaid.

Furthermore, there is no need for intermediaries. As a result, the agriculture industry and its clients can rest easy, confident that their needs will be satisfied (Xiong et al., 2020). Hungarian exchanges also offer blockchain technology. The stock markets can now leverage blockchain technology; however, their operations will remain the same. The futures market enables farmers to buy and sell crops, animals, vegetables, fruits, and other crops with adjusted prices. Farmers will understand their expenses, and customers will be prepared to predict price fluctuations (Kamilaris et al., 2019).

Retailers are needed to have all essential documents on file for food or other items purchases, even though the documents do not serve information about the health condition of the specific animal, its gender (cow or chicken), or the type of nutrition they received.

In the future, blockchain networks will enable applicants to save time when analyzing the history of acquired foodstuffs. These networks will present all related information about food production and transition that allow users to receive all pertinent details (Vishakha et al., 2021). The blockchain system that ensures the better permanence of all records enables swift and effortless access to a generally vast history of the food origins while on the move. This increases the probability that the food is both healthy and secure (Xu et al., 2022).

Walmart uses this contemporary technology to import livestock products from China and mangoes from Mexico. Walmart and IBM are partnering to implement blockchain technology in an agricultural supply chain aiming to enhance openness for retail customers (Astill et al., 2019).

The system deliberately seeks a centralized database on the blockchain network to gather data from all participants in the "food chain" (such as farmers, suppliers, and corporations). An increasing number of food and agricultural corporations are joining the system daily (Pham, 2018). IBM simultaneously partnered with ten other prominent food and consumer goods corporations worldwide to integrate blockchain technology into their supply chains. The agreement will enhance the food industry's ability to improve supply chain visibility and accountability.

According to Wang et al. (2019), companies such as Walmart, Nestle, Unilever, McCormick, Tyson, Kroger, McLane, Driscoll's, Dole, and Golden State Foods collectively generate more than $500 billion in global sales annually. In the foreseeable future, those businesses may start accepting Bitcoin and other digital currencies as a form of payment (IBM, 2019).

6.9 TRADITIONAL VERSUS BLOCKCHAIN METHODS USED IN AGRICULTURE

This chapter compares conventional methodologies with blockchain technology in agriculture segment. Conventional approaches such as manual record keeping, centralized databases, paper-based contracts, physical evidence for certification, and supply chain tracking have drawbacks such as inaccuracy, inefficiency, and vulnerability to detrimental activities.

Blockchain system offers more effective resolutions for supply chain monitoring, smart contracts for payouts and agreements, data management and sharing, certification and compliance, and finally financial inclusion and market access. This technology is transitioning agricultural practices by improving effectiveness and transparency.

6.9.1 TRADITIONAL METHODS USED IN AGRICULTURE DEPARTMENT CHALLENGES AND SOLUTIONS

Before introducing blockchain technology in agriculture sector, conventional approaches predominantly depended on paper-based methods or centralized databases for recording, tracking, and managing various agricultural activities (Altieri & Nicholls, 2017). Below are several conventional techniques that have been substituted for or enhanced by blockchain.

According to Tamayo (2020) in the past, farmers would keep track of their transactions, land ownership, crop yields, and sales by hand, using paper. This approach was susceptible to errors, loss, and tampering.

Paper-based contracts and transactions: farmers, suppliers, distributors, and retailers frequently recorded their agreements on physical paper which led to delays, disagreements, and inefficiencies throughout the supply chain (Marais, 2013).

We have used physical documentation to verify certifications like organic compatibility, fair trade, or sustainable farming techniques. However, it is probable that people could fake or alter these certificates (Maughan et al., 2016).

Tracing the trajectory of agricultural products from the farm to the table was difficult, depending only on documentation and confidence among the various stakeholders involved in the supply chain (Schuster, 2009).

6.9.2 Use of Blockchain Instead of Traditional Methods

The present utilization of blockchain methods in agriculture highlights their superiority over conventional approaches, offering enhanced transparency, efficiency, and traceability.

Blockchain facilitates the clear and unchangeable monitoring of agricultural products over every stage of the supply chain. The blockchain carefully documents every step of the transaction including planting, harvesting, processing, packing, and distribution.

This guarantees that consumers may track the source of their food goods, authenticate their genuineness, and assure adherence to quality and safety regulations (Salah et al., 2019). Blockchain offers immediate transparency, mitigates the risk of fraud and contamination, and strengthens consumer confidence alike conventional approaches that rely on paper-based records or centralized databases (Mirabelli & Solina, 2020).

Computer logic directly encodes the provisions of smart contracts to enable self-execution. These contracts in agriculture sector enable automated payments among farmers, suppliers, distributors, and retailers, contingent upon predetermined parameters like delivery milestones, quality benchmarks, or market prices (Pranto et al., 2021). This eliminates the need for intermediaries, lowers transactional funds, reduces controversy, and ensures time-oriented as well as transparent remittances. Unlike conventional paper contracts are vulnerable to errors, conflicts, and delays, smart contracts executed on the blockchain streamline operations and extend efficiency.

Blockchain enables safety and distributed storage of agricultural data namely crop production information, soil quality, weather trends, and farming methods. Blockchain system also allows farmers, researchers, and related corporations to securely swap and obtain data (Figure 6.2).

Data-driven decision-making, precision farming, and cooperative research are all made easier accordingly. As an alternative to centralized databases run by a single organization, farmers can retain ownership of their data on blockchain while letting authorized parties to access it, guaranteeing data privacy and sovereignty.

The process for acquiring certification concerning sustainable fair trade as well as organic farming methods could be made easier employing blockchain technology. It preserves a record of audits and certifications offering irrevocable proof of complying with legal and moral standards.

This method reduces the total loads of work related to administrative responsibilities, lowers the possibility of fraud, and increases the credibility of certification programs. On the other hand, traditional certification methods rely on tangible documents which can be challenging to manage, vulnerable to fraud and nearly impossible to confirm.

Block tech ensures small-scale farmers have access over financial services such as loans, insurance, and marketplaces without relying on conventional banking system, facilitating financial inclusion and market access. By taking advantage of such technology, creditors can record farmers' assets and transactions, which helps them gauge credibility, reduce risk, and offer tailored financial packages.

This promotes financial inclusion, serves small-scale farmers a greater voice, and strengthens rural small economies as well. Agricultural communities in remote or poorly developed regions are often too small for conventional financial institutions to operate efficiently. Employing blockchain-based resolutions that are more inclusive and accessible may address it.

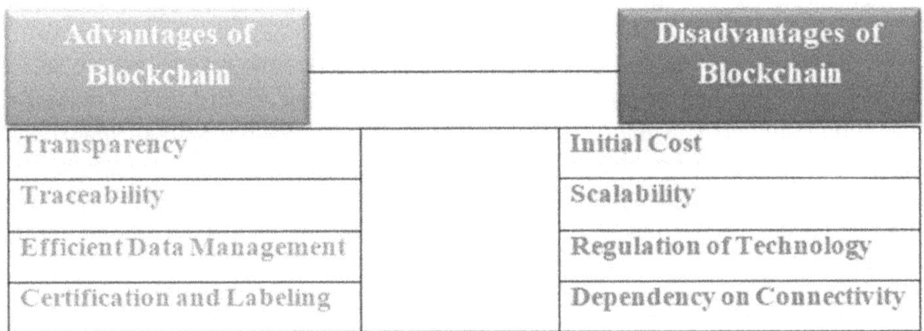

FIGURE 6.2   Advantages and disadvantages of blockchain in agriculture.

6.9.3 CHALLENGES WITH THE APPLICATION OF BLOCKCHAIN TECHNOLOGY IN THE AGRICULTURE SECTOR

Blockchain technology could lead to a remarkably important transition in agriculture sector. There are, however, a list of barriers and limits to be considered.

Using blockchain technology necessitates alarming financial resources for acquiring infrastructure components such as hardware, software, and highly skilled workforce. The initial expenses might be high for numerous farmers and small agricultural enterprises (Yadav & Singh, 2019).

Considering blockchain networks continuously grow in both size and complexity, vulnerabilities start to appear as a result. The enormous amount of data generated by the agriculture business segment may be far much for current blockchain platforms to handle, which could cause congestion and slow transaction processing (Thejaswini & Ranjitha, 2020).

Data privacy is a major challenge, even though blockchain network offers high security owing to its cryptographic methods. The probability of data breaches and unauthorized access could cause concern farmers as well as other stakeholders of disclosing sensitive information on a public blockchain.

The mispresent of interoperability among diverse blockchain platforms and current agricultural systems serves as a hurdle to its extensive acceptance and implementation. Integrating blockchain into current supply chain networks might pose challenges and incur expenses in the absence of norms and compiling procedures (Thejaswini & Ranjitha, 2020).

The agriculture segment of business is bound by a number of laws and guidelines tied to food safety, traceability, and sustainability to preserve regulatory compliance. Blockchain systems may need to be modified to comply with these criteria which vary according to the region and governmental authority. This requires careful planning and may add more complexity.

Throughout the entire agriculture industry, there are notable differences in terms of digital literacy and technological access among different individuals and areas. The digital gap may worsen in case farmers in distant or economically disadvantaged areas are unable to obtain the necessary infrastructure and knowledge for implementing blockchain technology (Yadav & Singh, 2019).

The energy-intensive character of blockchain validation, particularly in proof-of-work consensus systems such as the ones applied by Bitcoin, results to have raised in concerns regarding its potential environmental impact. According to Thejaswini and Ranjitha (2020), a thorough evaluation of the ecological effects of blockchain technology is essential as the agricultural sector attempts to improve sustainability.

Like any new technology, stakeholders who were used to traditional business practices may oppose blockchain. Due to their lack of expertise, mistrust, or aversion to change, farmers and suppliers that are involved in the agriculture supply chain may be unwilling to adopt blockchain technology (Yadav & Singh, 2019).

Legislators, technology developers, and agricultural stakeholders are needed to develop blockchain-based alternatives that are secure, scalable, and ubiquitous in order to address such issues. It is important that we specifically design these resolutions to fulfill the particular demands of the agricultural industry.

6.9.4 Solution to Overcome Blockchain Challenges in Agriculture

A comprehensive strategy encompassing technological advancements, legislative interventions, and industry stakeholder collaboration is really necessary to overcome the restrictions of blockchain technology in the agriculture sector. The following are a few possible resolutions.

Employing cost-effective blockchain technologies such as cloud-based operating platforms or collaborative networks could diminish the initial financial commitment required for implementation. Government agencies and industry associations can offer financial assistance like grants, subsidies, or tax advantages, to assist farmers and also small businesses in implementing blockchain technology (Jothikumar, 2021).

The main objective of R&D should be to make better blockchain technology's innovative techniques such as sharding, side chains, and layer-2 solutions such as state channels and plasma. Moreover, enhancing efficiency of the smart contract execution and also transaction processing algorithms can boost network accomplishment.

Improved data security and privacy: On public blockchains, privacy-preserving techniques like zero-knowledge proofs, homomorphic security encryption as well as safe multi-party computation can be used to protect privacy. In addition, the accomplishment of robust access control methods and encryption guidelines can effectively safeguard sensitive data.

Establishing global interoperability protocols and standards can enable smooth and efficient interchange as well as integration of data between diverse blockchain technology platforms and existing agricultural systems. Collaborative endeavors, like industrial consortia or software development communities, be able to contribute to the advancement of standardization efforts.

Governments as well as regulatory authorities, in partnership with industry players, can establish regulatory frameworks to offer explicit guidelines and regulations for implementing blockchain technology in the agriculture sector. Promoting safe blockchain usage requires compliance with both existing regulations and encouraging innovation and flexibility (Jothikumar, 2021).

By allocating resources to digital infrastructure, training programs, and community engagement projects, we are able to narrow the gap among access to technology and enable farmers in underserved regions to effectively use blockchain technology. Public–private collaborations as well as grassroots efforts are essential to advancing digital literacy and ensuring widespread access to technology (Ferrag et al., 2020).

Promoting the implementation of energy-efficient consensus processes such as PoS or delegated PoS can helpfully decrease the environmental effect of blockchain network. In addition, research on alternate consensus algorithms and renewable energy sources can assist in further decreasing environmental problems (Ferrag et al., 2020).

Increasing the knowledge and consciousness of blockchain technology among farmers, suppliers, and other stakeholders is vital so as to overcome opposition to change. Education and training programs, professional conferences as well as knowledge-sharing platforms can clarify the intricacies of blockchain and show its potential advantages (Jothikumar, 2021).

Through cooperative endeavors and inventive approaches, the agricultural sector can use the revolutionary capabilities of blockchain technology to enhance visibility and productivity as well as traceability along the whole supply chain.

6.10   DISCUSSIONS

After conducting a large-scale analysis of the literature on blockchain technology in agriculture and food systems, it is clear that blockchain be able to provide substantial benefits and answers to long-standing difficulties encountered by common approaches. The debate's initial focus revolved around the integration of blockchain technology in Germany's agricultural sector. This gives a demonstration of the capacity of blockchain to improve coordination mechanisms in vertically integrated food supply chains and provides useful knowledge about its appropriateness and effectiveness in managing supply chain activities and fostering collaboration among stakeholders. Sections 2 and 3 examined the application of blockchain technology in the agricultural sectors of Pakistan and Hungary, respectively. The focus was on highlighting how blockchain may enhance transparency, traceability as well as efficiency across the whole agricultural value chain. In addition, parts 4, 5, and 6 give a more thorough examination of the advantages and difficulties linked to the implementation of blockchain technology in the agricultural sector. The highlighted parts emphasize the capacity of blockchain technology to transform supply chain management, improve food security, reduce environmental risks, and inspire sustainable farming methods. Nevertheless, they also highlighted other snags such as the ecological viability of blockchain technology, the necessity for effective consensus algorithms, and the need to tackle concerns over data privacy and security.

In addition, Section 7 provides a comparison between conventional agricultural methods and blockchain-based models that highlight transformative potential of blockchain in some agricultural domains. Also, conventional methods are inefficient, while blockchain provides solutions for tracking, automated agreements, data control, as well as financial access. Collaborative efforts, technical developments, and legal frameworks are necessary to solve all challenges regarding high costs, scaling concerns as well as opposition to change. By overcoming these challenges, this sector may use blockchain technology's revolutionary potential to creating a more environmentally convivial food system which is open, accountable, and hold out against adversity.

We conducted an extensive analysis to examine the current global implementation of blockchain technology in agriculture and food systems, along with its benefits, challenges, and potential applications in enhancing farming operations, supply chain management, food safety as well as security. An analysis of multiple case studies from various nations, such as South America, Germany, Pakistan, and Hungary, earned significant insights into the present status of blockchain deployment in distinct agricultural environments. Comparing traditional methods with blockchain solutions showed how revolutionary blockchain is a way to solve long-standing problems like data management, supply chain traceability, and certification processes. This study focuses on these objectives that provide significant contributions to both academia and industry and facilitating future research and practical applications in the agricultural technology segment.

6.11 SUMMARY

The thorough examination of blockchain technology adoption in agriculture and food sector demonstrates how revolutionary it might be for supply chain tracking, agricultural operations management, and tracking food goods from farm to plate.

Blockchain provides irreversible, transparent as well as decentralized platforms for data management and transaction follow-up. It provides resolutions to resistant issues over agricultural sector, including fraud, inefficiencies, and lack of confidence.

Blockchain network also can improve cooperation among stakeholders, empower supply chain openness, guarantee food safety and security, encourage sustainable farming practices and at the end, facilitate financial inclusion for small-scale rated farmers.

Various developing and non-developing countries, such as the United States, Germany, Pakistan, and Hungary, have witnessed these benefits. However, this study aims for a comprehensive strategy for harnessing the full potential of blockchain technology in the agricultural sector.

Investment in extra research and development is notable for understanding the effects of such investments on different types of farms and assessing their financial consequences. Effective communication among stakeholders is necessary for successfully overcoming burdens and fostering innovation. Specifically designed for the agriculture industry, personalized integration ensures compatibility, scalability, and cost-efficiency. Prioritizing affordability and accessibility are valuable especially for farms of smaller sizes. In order to manage growing transaction volumes and enable seamless data sharing, flexibility, and compatibility ought to be prioritized.

To promote awareness and acknowledgment among farmers as well as stakeholders, education and training programs are absolutely essential. By considering these recommendations, agricultural sector may promote a more resilient and sustainable food system via using blockchain tech to rocket transparency, traceability, and efficiency.

Subsequent research on blockchain technology about agriculture business sector is required to concentrate on resolving the remarkable constraints as well as exploring the potential benefits and uses of blockchain in diverse agricultural contexts. Doing research remains required to develop more energy-efficient consensus algorithms, enrich the scalability and smooth operation of blockchain technology, and adjust regulatory frameworks for vast use of blockchain while lowering the cost of deployment.

It is also pretty important to continue researching how blockchain technology may affect small-scale farmers, rural communities, and global food chains from a socioeconomic standpoint. Furthermore, it is imperative to establish measures that facilitate the integration of blockchain technology with conventional agricultural methods through comprehensive training and capacity development for stakeholders at all levels. To fully use the benefits of blockchain technology, agriculture sector should take into account these mentioned future factors. It will enable the creation of a food system which could be more transparent, efficient, and sustainable.

REFERENCES

Alkahtani, M., Khalid, Q.S., Jalees, M., Omair, M., Hussain, G. and Pruncu, C.I., 2021. E-agricultural supply chain management coupled with blockchain effect and cooperative strategies. *Sustainability*, *13*(2), p.816.

Alobid, M., Abujudeh, S. and Szűcs, I., 2022. The role of blockchain in revolutionizing the agricultural sector. *Sustainability*, *14*(7), p.4313. https://doi.org/10.3390/su14074313.

Altieri, M.A. and Nicholls, C.I., 2017. The adaptation and mitigation potential of traditional agriculture in a changing climate. *Climatic Change*, *140*, pp.33–45.

Ante, L., Sandner, P. and Fiedler, I., 2018. Blockchain-based ICOs: Pure hype or the dawn of a new era of startup financing? *Journal of Risk and Financial Management*, *11*(4), p.80.

Antonucci, F., Figorilli, S., Costa, C., Pallottino, F., Raso, L. and Menesatti, P., 2019. A review on blockchain applications in the agri-food sector. *Journal of the Science of Food and Agriculture*, *99*(14), pp.6129–6138.

Astill, J., Dara, R.A., Campbell, M., Farber, J.M., Fraser, E.D., Sharif, S. and Yada, R.Y., 2019. Transparency in food supply chains: A review of enabling technology solutions. *Trends in Food Science & Technology*, *91*, pp.240–247.

Awan, S.H., Ahmed, S., Safwan, N., Najam, Z., Hashim, M.Z. and Safdar, T., 2019. Role of internet of things (IoT) with blockchain technology for the development of smart farming. *Journal of Mechanics of Continua and Mathematical Sciences*, *14*(5), pp.170–188.

Awan, S.H., Ahmed, S., Nawaz, A., Sulaiman, S., Zaman, K., Ali, M.Y., Najam, Z. and Imran, S., 2020a. BlockChain with IoT, an emergent routing scheme for smart agriculture. *International Journal of Advanced Computer Science and Applications*, *11*(4), pp.420–429. https://doi.org/10.14569/ijacsa.2020.0110457.

Awan, S.H., Ahmad, S., Khan, Y., Safwan, N., Qurashi, S.S. and Hashim, M.Z., 2021. A combo smart model of blockchain with the Internet of Things (IoT) for the transformation of agriculture sector. *Wireless Personal Communications*, *121*(3), pp.2233–2249.

Awan, S.H., Nawaz, A., Ahmed, S., Khattak, H.A., Zaman, K. and Najam, Z., 2020b. Blockchain based smart model for agricultural food supply chain. In *2020 International Conference on UK-China Emerging Technologies (UCET)* [Preprint]. https://doi.org/10.1109/ucet51115.2020.9205477.

Behnke, K. and Janssen, M.F.W.H.A., 2020. Boundary conditions for traceability in food supply chains using blockchain technology. *International Journal of Information Management*, *52*, p.101969.

Bermeo-Almeida, O., Cardenas-Rodriguez, M., Samaniego-Cobo, T., Ferruzola-Gómez, E., Cabezas-Cabezas, R. and Bazán-Vera, W., 2018. Blockchain in agriculture: A systematic literature review. In *Technologies and Innovation: 4th International Conference, CITI 2018, Guayaquil, Ecuador, November 6–9, 2018, Proceedings 4* (pp.44–56). Springer International Publishing.

Blanchard, D., 2021. *Supply chain management best practices.* John Wiley & Sons.

Casino, F., Dasaklis, T.K. and Patsakis, C., 2019. A systematic literature review of blockchain-based applications: Current status, classification and open issues. *Telematics and Informatics*, *36*, pp.55–81.

Chofreh, A.G., Goni, F.A., Malik, M.N., Khan, H.H. and Klemeš, J.J., 2019. The imperative and research directions of sustainable project management. *Journal of Cleaner Production*, *238*, p.117810.

Chohan, U.W., 2019. *Initial coin offerings (ICOs): Risks, regulation and accountability* (pp.165–177). Springer International Publishing.

Davidson, S., De Filippi, P. and Potts, J., 2018. Blockchains and the economic institutions of capitalism. *Journal of Institutional Economics*, *14*(4), pp.639–658.

Deloitte, 2020. *2020 Global blockchain survey.* [Online]. Available: www2.deloitte.com/content/dam/insights/us/articles/6608_2020-global-blockchain-survey/DI_CIR%20 2020%20global%20blockchain%20survey.pdf. [Accessed: 27 December 2020].

Demestichas, K., Peppes, N., Alexakis, T. and Adamopoulou, E., 2020. Blockchain in agriculture traceability systems: A review. *Applied Sciences*, *10*(12), p.4113.

De Vries, A., 2020. Bitcoin's energy consumption is underestimated: A market dynamics approach. *Energy Research & Social Science*, *70*, p.101721.

Dey, K. and Shekhawat, U., 2021. Blockchain for sustainable e-agriculture: Literature review, architecture for data management and implications. *Journal of Cleaner Production*, *316*, p.128254.

Duan, J., Zhang, C., Gong, Y., Brown, S. and Li, Z., 2020. A content-analysis based literature review in blockchain adoption within food supply chain. *International Journal of Environmental Research and Public Health*, *17*(5), p.1784.

Dujak, D. and Sajter, D., 2018. Blockchain applications in supply chain. In *Ecoproduction* (pp.21–46). Springer, Cham, https://doi.org/10.1007/978-3-319-91668-2_2.

Ehsan, I., Irfan Khalid, M., Ricci, L., Iqbal, J., Alabrah, A., Sajid Ullah, S. and Alfakih, T.M., 2022. A conceptual model for blockchain-based agriculture food supply chain system. *Scientific Programming*, *2022*, pp.1–15.

Elijah, O., Rahman, T.A., Orikumhi, I., Leow, C.Y. and Hindia, M.N., 2018. An overview of Internet of Things (IoT) and data analytics in agriculture: Benefits and challenges. *IEEE Internet of things Journal*, 5(5), pp.3758–3773.

Farooq, M.S., Ansari, Z.K., Alvi, A., Rustam, F., Díez, I.D.L.T., Mazón, J.L.V., Rodríguez, C.L. and Ashraf, I., 2024. Blockchain based transparent and reliable framework for wheat crop supply chain. *PLoS One*, 19(1), p.e0295036.

Farooq, M.S., Riaz, S., Rehman, I.U., Khan, M.A. and Hassan, B., 2023. A blockchain-based framework to make the rice crop supply chain transparent and reliable in agriculture. *Systems*, 11(9), p.476.

Ferrag, M.A., Shu, L., Yang, X., Derhab, A. and Maglaras, L., 2020. Security and privacy for green IoT-based agriculture: Review, blockchain solutions and challenges. *IEEE Access*, 8, pp.32031–32053.

Galvez, J.F., Mejuto, J.C. and Simal-Gandara, J., 2018. Future challenges on the use of blockchain for food traceability analysis. *TrAC Trends in Analytical Chemistry*, 107, pp.222–232.

Ganne, E., 2018. *Can blockchain revolutionize international trade?* (p.152). Geneva: World Trade Organization.

Ge, L., Brewster, C., Spek, J., Smeenk, A., Top, J., Van Diepen, F., Klaase, B., Graumans, C. and de Wildt, M.D.R., 2017. *Blockchain for agriculture and food: Findings from the pilot study* (No. 2017-112). Wageningen Economic Research.

Groombridge, D., 2020. *Unpacking blockchain myths from the reality*. Gartner Webinars.

Gurtu, A. and Johny, J., 2019. Potential of blockchain technology in supply chain management: A literature review. *International Journal of Physical Distribution & Logistics Management*, 49(9), pp.881–900.

Hakim, A., Shafique, S., Mehdi, M. and Baig, I.A., 2024. Farmer to consumer – an online supply chain process system using blockchain technology. *Agricultural Sciences Journal*, 6(1), pp.1–11.

Hanf, J. and Dautzenberg, K., 2006. A theoretical framework of chain management. *Journal on Chain and Network Science*, 6(2), pp.79–94.

Huh, J.H. and Kim, S.K., 2019. The blockchain consensus algorithm for viable management of new and renewable energies. *Sustainability*, 11(11), p.3184.

IBM, 2019. *IBM food trust – a new era for the world's food supply*. [Online]. Available: www.ibm.com/blockchain/solutions/food-trust. [Accessed: 31 August 2019].

Jothikumar, R., 2021. Applying blockchain in agriculture: A study on blockchain technology, benefits and challenges. *Deep learning and edge computing solutions for high performance computing* (pp.167–181). Cham: Springer.

Jović, M., Tijan, E., Žgaljić, D. and Aksentijević, S., 2020. Improving maritime transport sustainability using blockchain-based information exchange. *Sustainability*, 12(21), p.8866.

Kamble, S.S., Gunasekaran, A. and Gawankar, S.A., 2020. Achieving sustainable performance in a data-driven agriculture supply chain: A review for research and applications. *International Journal of Production Economics*, 219, pp.179–194.

Kamilaris, A., Fonts, A. and Prenafeta-Boldú, F.X., 2019. The rise of blockchain technology in agriculture and food supply chains. *Trends in Food Science & Technology*, 91, pp.640–652.

Khalil, M.M. and Ahmed, W., 2024. Analyzing the drivers of blockchain adoption for supply chain in Pakistan. *Journal of Science and Technology Policy Management [Preprint]*. https://doi.org/10.1108/jstpm-10-2023-0178.

Khan, A.A., Shaikh, Z.A., Belinskaja, L., Baitenova, L., Vlasova, Y., Gerzelieva, Z., Laghari, A.A., Abro, A.A. and Barykin, S., 2022. A blockchain and metaheuristic-enabled distributed architecture for smart agricultural analysis and ledger preservation solution: A collaborative approach. *Applied Sciences*, 12(3), p.1487.

Khan, H.H., Malik, M.N., Konečná, Z., Chofreh, A.G., Goni, F.A. and Klemeš, J.J., 2022. Blockchain technology for agricultural supply chains during the COVID-19 pandemic: Benefits and cleaner solutions. *Journal of Cleaner Production, 347*, p.131268.

Kouhizadeh, M. and Sarkis, J., 2018. Blockchain practices, potentials and perspectives in greening supply chains. *Sustainability, 10*(10), p.3652.

Kramer, M.P., Bitsch, L. and Hanf, J., 2021. Blockchain and its impacts on agri-food supply chain network management. *Sustainability, 13*(4), p.2168.

Li, G., Chen, D., Zhang, J., Hu, F. and Zheng, G., 2021. Construction of simplified traceability system of agricultural products based on android. In *IOP Conference Series: Earth and Environmental Science*. IOP Publishing Ltd.

Liu, W., Shao, X.F., Wu, C.H. and Qiao, P., 2021. A systematic literature review on applications of information and communication technologies and blockchain technologies for precision agriculture development. *Journal of Cleaner Production, 298*, p.126763.

Marais, A., 2013. *Developing a managerial framework for e-contracting in the agricultural business environment* (Doctoral dissertation).

Marzougui, F., Elleuch, M. and Kherallah, M., 2023, December. Literature review of IoT and blockchain technology in agriculture. In *2023 24th International Arab Conference on Information Technology (ACIT)* (pp.1–8). IEEE.

Maughan, T., Drost, D., Olsen, S. and Black, B., 2016. *Good agricultural practices (GAP): Certification basics.*

Mba, C., Abang, M., Diulgheroff, S., Hrushka, A., Hugo, W., Ingelbrecht, I., Jankuloski, L., Leskien, D., Lopez, V., Muminjanov, H., Mulila Mitti, J., Nersisyan, A., Noorani, A., Piao, Y. and Sagnia, S., 2020. FAO supports countries in the implementation of the second global plan of action for plant genetic resources for food and agriculture. In *International Symposium on Survey of Uses of Plant Genetic Resources to the Benefit of Local Populations 1267* (pp.197–208). ISHS Acta Horticulturae, https://doi.org/10.17660/ActaHortic.2020.1267.30.

Mirabelli, G. and Solina, V., 2020. Blockchain and agricultural supply chains traceability: Research trends and future challenges. *Procedia Manufacturing, 42*, pp.414–421.

Motta, G.A., Tekinerdogan, B. and Athanasiadis, I.N., 2020. Blockchain applications in the agri-food domain: The first wave. *Frontiers in Blockchain, 3*, p.6.

Mukherjee, A.A., Singh, R.K., Mishra, R. and Bag, S., 2022. Application of blockchain technology for sustainability development in agricultural supply chain: Justification framework. *Operations Management Research, 15*(1), pp.46–61.

Ordóñez, J., Alexopoulos, A., Koutras, K., Kalogeras, A., Stefanidis, K. and Martos, V., 2023. Blockchain in agriculture: A PESTELS analysis. *IEEE Access, 11*, pp.73647–73679. https://doi.org/10.1109/access.2023.3295889.

Pham, H., 2018. *The impact of blockchain technology on the improvement of food supply chain management: Transparency and traceability: A case study of Walmart and Atria.*

Pranto, T.H., Noman, A.A., Mahmud, A. and Haque, A.B., 2021. Blockchain and smart contract for IoT enabled smart agriculture. *PeerJ Computer Science, 7*, p.e407.

Priyadarshini, I., 2019. Introduction to blockchain technology. *Cyber security in parallel and distributed computing: Concepts, techniques, applications and case studies* (pp.91–107). Wiley eBooks. https://doi.org/10.1002/9781119488330.

Reyes, S.R., Miyazaki, A., Yiu, E. and Saito, O., 2020. Enhancing sustainability in traditional agriculture: Indicators for monitoring the conservation of Globally Important Agricultural Heritage Systems (GIAHS) in Japan. *Sustainability, 12*(14), p.5656.

Rocha, G.D.S.R., de Oliveira, L. and Talamini, E., 2021. Blockchain applications in agribusiness: A systematic review. *Future Internet, 13*(4), p.95.

Salah, K., Nizamuddin, N., Jayaraman, R. and Omar, M., 2019. Blockchain-based soybean traceability in agricultural supply chain. *IEEE Access*, 7, pp.73295–73305.

Schuster, E.W., 2009. Agricultural supply chains: Track and trace for improved food safety. *Acta Horticulturae*, 824, pp.113–120. https://doi.org/10.17660/actahortic.2009. 824.12.

Sendros, A., Drosatos, G., Efraimidis, P.S. and Tsirliganis, N.C., 2022. Blockchain applications in agriculture: A scoping review. *Applied Sciences*, 12(16), p.8061.

Sharma, P., Jindal, R. and Borah, M.D., 2022. A review of smart contract-based platforms, applications and challenges. *Cluster Computing*, 26(1), pp.395–421.

Sharma, R., Kamble, S.S., Gunasekaran, A., Kumar, V. and Kumar, A., 2020. A systematic literature review on machine learning applications for sustainable agriculture supply chain performance. *Computers & Operations Research*, 119, p.104926.

Sharma, R., Samad, T.A., Jabbour, C.J.C. and de Queiroz, M.J., 2021. Leveraging blockchain technology for circularity in agricultural supply chains: Evidence from a fast-growing economy. *Journal of Enterprise Information Management* [Preprint]. https://doi.org/ 10.1108/jeim-02-2021-0094.

Song, C. and Li, C., 2021. Research on agricultural products supply chain traceability system: Blockchain consensus algorithm optimization. In *2021 9th International Conference on Traffic and Logistic Engineering, ICTLE 2021* (pp.64–68). Institute of Electrical and Electronics Engineers Inc.

Song, L., Wang, X., Wei, P., Lu, Z., Wang, X. and Merveille, N., 2021. Blockchain-based flexible double-chain architecture and performance optimization for better sustainability in agriculture. *Computers, Materials & Continua*, 68, pp.1429–1446.

Tamayo, J., 2020. Development of agriculture office farmers' record management system for the municipality of Mangaldan. *Southeast Asian Journal of Science and Technology*, 5(1).

Thejaswini, S. and Ranjitha, K.R., 2020, January. Blockchain in agriculture by using decentralized peer to peer networks. In *2020 Fourth International Conference on Inventive Systems and Control (ICISC)* (pp.600–606). IEEE.

Treiblmaier, H., 2018. The impact of the blockchain on the supply chain: A theory-based research framework and a call for action. *Supply Chain Management: An International Journal*, 23(6), pp.545–559.

Treiblmaier, H., 2019. Toward more rigorous blockchain research: Recommendations for writing blockchain case studies. *Frontiers in Blockchain*, 2. https://doi.org/10.3389/ fbloc.2019.00003

Ullah, N., 2021, December. Blockchain technology in smart agriculture environment: A PLS-SEM. In *2021 International Conference on Electronic Information Technology and Smart Agriculture (ICEITSA)* (pp.514–519). IEEE.

Vaia, G., 2020. *Blockchain technology in meat supply chain: Operational and economic implications*. http://hdl.handle.net/10579/17196.

Vangala, A., Das, A.K., Kumar, N. and Alazab, M., 2020. Smart secure sensing for IoT-based agriculture: Blockchain perspective. *IEEE Sensors Journal*, 21(16), pp.17591–17607.

Van-Wassenaer, L., Verdouw, C. and Wolfert, S., 2021. What blockchain are we talking about? An analytical framework for understanding blockchain applications in agriculture and food. *Frontiers in Blockchain*, 4, p.653128.

Varriale, V., Cammarano, A., Michelino, F. and Caputo, M., 2020. The unknown potential of blockchain for sustainable supply chains. *Sustainability*, 12(22), p.9400.

Vishakha, B.S., Sharma, N., Bhushan, B. and Kaushik, I., 2021. Blockchain-based cultivating ideas for growth: A new agronomics perspective. In *Advances in computing communications and informatics* (pp.195–219). Bentham Science Publishers eBooks. https://doi. org/10.2174/9781681088624121010012

Vranken, H., 2017. Sustainability of Bitcoin and blockchains. *Current Opinion in Environmental Sustainability*, *28*, pp.1–9.

Wang, Y., Han, J.H. and Beynon-Davies, P., 2019. Understanding blockchain technology for future supply chains: A systematic literature review and research agenda. *Supply Chain Management: An International Journal*, *24*(1), pp.62–84.

Westerkamp, M., Victor, F. and Küpper, A., 2020. Tracing manufacturing processes using blockchain-based token compositions. *Digital Communications and Networks*, *6*(2), pp.167–176.

Xiong, H., Dalhaus, T., Wang, P. and Huang, J., 2020. Blockchain technology for agriculture: Applications and rationale. *Frontiers in Blockchain*, *3*, p.7.

Xu, Y., Li, X., Zeng, X., Cao, J. and Jiang, W., 2022. Application of blockchain technology in food safety control: Current trends and future prospects. *Critical Reviews in Food Science and Nutrition*, *62*(10), pp.2800–2819.

Yadav, V.S. and Singh, A.R., 2019, September. Use of blockchain to solve select issues of Indian farmers. In *AIP Conference Proceedings* (Vol. 2148, No. 1). AIP Publishing.

Zhang, R., Xue, R. and Liu, L., 2019. Security and privacy on blockchain. *ACM Computing Surveys (CSUR)*, *52*(3), pp.1–34.

Zhu, Q. and Kouhizadeh, M., 2019. Blockchain technology, supply chain information and strategic product deletion management. *IEEE Engineering Management Review*, *47*(1), pp.36–44.

7 Blockchain and Circular Economy

Hamed Taherdoost and Mitra Madanchian

7.1 INTRODUCTION

Integrating blockchain technology with the circular economy offers a revolutionary method for implementing sustainable economic practices (Yildizbasi, 2021). This chapter examines the interconnections between blockchain technology and the circular economy, investigating how blockchain can transform resource management, supply chains, and sustainability efforts.

The circular economy is a progressive economic paradigm that seeks to eliminate waste and optimize the value of resources through the promotion of reuse, repair, and recycling. In contrast to the conventional linear economy that adheres to a "take-make-dispose" model, the circular economy prioritizes the ongoing utilization of resources inside a closed-loop system to minimize environmental harm and promote sustainable progress (Lazarevic & Brandão, 2020). Blockchain technology is a distributed ledger system that securely records transactions over a network of computers. It is recognized for its decentralized and transparent nature. Blockchain technology improves trust, efficiency, and accountability in different industries by providing secure and tamper-proof data storage and transparent transaction tracking (Xu et al., 2020).

Implementing blockchain technology in the circular economy provides a multitude of advantages, such as improved traceability, visibility, and effectiveness in resource allocation. Blockchain technology permits the monitoring of items from their inception to their end, encourages environmentally friendly activities, and allows for the development of inventive methods to reduce waste and optimize resources. The immutability and decentralized nature of the system create a strong basis for constructing a circular economy ecosystem that is more sustainable and resilient (Dapp, 2018).

7.2 UNDERSTANDING CIRCULAR ECONOMY PRINCIPLES

Within the context of the circular economy, the ideas of "Reduce, Reuse, Recycle" serve as fundamental principles that direct the sustainable management of resources and the minimization of waste. These principles are essential for transforming conventional linear economic models into a more circular and ecologically responsible approach (Madaan et al., 2024).

The idea of reduction emphasizes the importance of reducing resource consumption and minimizing waste output. Through the reduction of superfluous consumption

DOI: 10.1201/9781003609865-7

and the efficient usage of resources, both individuals and enterprises have the ability to greatly diminish their impact on the environment and actively contribute to the advancement of a more sustainable economy. Reuse, which is another fundamental principle, centers on prolonging the durability of items and materials by discovering inventive methods to repurpose them. Embracing reuse includes actions such as refurbishing, mending, and sharing products to avoid premature destruction. This method not only preserves resources but also minimizes energy usage and redirects a significant quantity of garbage away from landfills. Recycling, as the third principle, has a crucial part in the circular economy as it transforms waste materials into new goods or raw materials. The circular economy seeks to achieve resource efficiency, reduce dependence on new resources, and address environmental degradation by implementing recycling methods for items such as plastics, metals, and paper (Bucknall, 2020; de Sa & Korinek, 2021).

The supplied sources offer unique perspectives on closed-loop systems and their importance in waste management and resource preservation. These systems have the objective of diminishing, reutilizing, and recycling materials and energy in order to limit the negative effects on the environment and foster sustainability. Closed-loop recycling, in contrast to open-loop recycling, aims to convert goods or materials into new ones without requiring more raw materials, hence minimizing environmental damage and avoiding resource exhaustion (Kara et al., 2022).

When it comes to managing plant waste, designing a closed-loop system requires identifying and measuring various types of waste, assessing the waste hierarchy, implementing strategies to reduce waste, and monitoring waste performance to improve efficiency, sustainability, and environmental impact. The concepts of "Reduce, Reuse, Recycle" are essential in closed-loop systems, prioritizing waste reduction, minimization, reuse, recycling, and responsible disposal.

Closed-loop systems are essential for environmental sustainability since they help conserve natural resources, minimize waste, decrease greenhouse gas emissions, and enhance economic efficiency (Cobîrzan et al., 2023). Industries can diminish their landfill contributions, conserve space, and decrease production costs by recycling materials such as aluminum, glass, and plastic inside closed-loop systems. By implementing closed-loop techniques, such as the practice of composting organic waste, it is possible to achieve a substantial reduction in carbon dioxide emissions and make a valuable contribution to mitigating climate change (Dsouza et al., 2021).

Implementing measures to extend the lifespan of products is essential for shifting toward a more environmentally friendly and circular economy. The fundamental objective of these tactics is to consistently provide assistance and safeguard a product throughout its entire life cycle, ensuring it remains in the mature stage and preventing it from entering the decline phase. By doing this, the amount of value obtained from the product is maximized before it is eventually thrown away, which reduces the requirement for additional, unused resources (Van Ewijk & Stegemann, 2016).

Efficient ways for extending the lifespan of a product involve a variety of different methods. Businesses can distinguish their products by emphasizing distinctive attributes, functionality, and advantages, revamping and rejuvenating the product to expand its appeal, and introducing novel sizing or customization choices. In addition, offering repair, refurbishment, and remanufacturing services, together with facilitating secondary markets and online resale platforms, can greatly prolong the operational lifespan of a device (Park, 2009).

Adopting these tactics for extending the lifespan of products can be advantageous for both businesses and customers. For corporations, it has the potential to decrease expenses and improve their commitment to corporate social responsibility. For customers, it offers items that are more resilient and of superior quality, resulting in long-term cost savings, despite the possibility of higher initial expenses. Crucial for enabling product life extension and the circular economy, closed-loop systems play a vital role in recycling materials and minimizing waste. These systems ensure the continual reuse and repurposing of resources (Mestre & Cooper, 2017).

7.3 CHALLENGES IN IMPLEMENTING CIRCULAR ECONOMY

The obstacles to adopting the circular economy, including the absence of traceability in supply chains, are substantial impediments that must be resolved in order to promote a more sustainable economic model (Figure 7.1). Traceability is essential for maintaining openness, accountability, and efficiency in supply chains, particularly for monitoring the movement of materials, products, and resources over their entire lifespan (Garcia-Torres et al., 2019). Lack of sufficient traceability methods poses difficulties in monitoring and verifying the source, quality, and sustainability of resources, impeding endeavors to foster circularity and minimize waste (Leng et al., 2020).

The absence of traceability in supply chains presents numerous challenges to the effective implementation of the circular economy. It might result in challenges when trying to determine the origins of products, evaluate their environmental consequences, and guarantee adherence to sustainability criteria. In the absence of adequate traceability, the capacity to track the flow of items for repair, refurbishment, or recycling is compromised, hence reducing the efficiency of closed-loop systems and resource optimization tactics (Bressanelli et al., 2019).

To tackle the problem of traceability in supply chains, it is necessary for stakeholders from various industries, legislators, and technology suppliers to work together in a collaborative manner (Bhatt et al., 2016). By using cutting-edge technologies such as blockchain, IoT, and data analytics, traceability may be significantly improved. This is achieved through the creation of transparent and unchangeable records of transactions and product movements. Enhancing traceability in supply chains enables organizations to more effectively monitor the origin of goods, improve resource utilization, minimize waste, and establish a more robust basis for a sustainable circular economy (Feng et al., 2020).

The inefficient allocation of resources presents a major obstacle in the implementation of the circular economy, affecting all facets of resource management and sustainability. A significant factor that leads to inefficiencies is the absence of traceability in supply chains. Lack of reliable traceability mechanisms hinders the ability to effectively monitor the source, quality, and sustainability of materials, thereby obstructing efforts to encourage circularity and minimize waste. The absence of transparency impedes the efficient monitoring of materials throughout their entire lifespan, restricting the ability to optimize resource utilization and impeding the creation of closed-loop systems (Liu et al., 2018).

Contamination and additive levels in waste streams pose an additional barrier to effective resource allocation. The presence of hazardous or poisonous compounds in some design aspects makes recycling procedures more difficult and costly, resulting in material losses and limiting the ability to effectively recover resources. Furthermore, problems like as downcycling and material deterioration, which are especially common in plastic recycling, lead to a restricted number of recycling cycles and a significant proportion of downcycled materials. This has a negative impact on the overall effectiveness of resource use (Xiong et al., 2019).

Moreover, the absence of economically viable recycling technology and the limited market demand for recyclable materials pose obstacles to achieving optimal resource allocation. Technological constraints and the lack of practical recycling methods for certain materials lead to inefficiencies in the recovery and exploitation of resources. The difficulty of inefficient resource allocation, particularly in handling plastic packaging trash and other waste streams, is worsened by inadequate waste collection systems and economic considerations such as low prices of virgin materials and high costs of sorting and processing (Fan et al., 2020).

To tackle these difficulties, a comprehensive strategy is needed. This strategy should involve strengthening the ability to track products in supply chains, encouraging the adoption of sustainable design practices, investing in cutting-edge recycling technologies, and enhancing garbage collection infrastructure. By surmounting these challenges and employing tactics to maximize resource distribution, firms and industries can make progress toward achieving the tenets of the circular economy and promoting a more sustainable and efficient future (Iacovidou et al., 2021).

Transparency is crucial for improving traceability in supply chains, guaranteeing the smooth movement of commodities and resources, and fostering sustainable practices. Trust is crucial in promoting collaboration among stakeholders, facilitating the convergence of interests, and establishing a solid basis for effective partnerships within the circular economy framework.

The absence of openness and trust can impede the transition to a circular economy by obstructing the exchange of information, restraining collaboration, and erecting obstacles to the adoption of sustainable practices. Within the realm of supply chains, it is imperative to have improved visibility and ability to track products in order to remove barriers to circular commerce, foster equitable and inclusive practices, and tackle environmental and social issues related to resource management (Anbumozhi & Kimura, 2018).

In order to tackle these difficulties, it is essential for stakeholders to give priority to establishing trust and transparency along the whole value chain. This entails the exchange of data regarding the composition of materials, techniques of production, and behaviors related to sustainability in order to optimize the use of resources and encourage circularity. By cultivating an environment characterized by openness and confidence, entrepreneurs, politicians, and other participants can collaborate to establish a circular economy ecosystem that is more enduring and environmentally friendly, ultimately benefiting society as a whole (Nogueira et al., 2020).

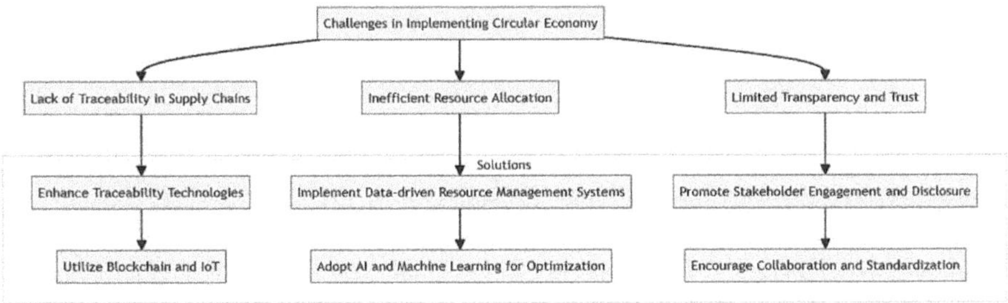

FIGURE 7.1 Examining obstacles and resolutions in the implementation of circular economy.

7.4 ROLE OF BLOCKCHAIN IN CIRCULAR ECONOMY

The primary function of blockchain technology in the circular economy is to improve traceability and transparency, which are essential components in the shift toward a more sustainable and efficient economic model (Centobelli et al., 2022).

The transparency and immutability of blockchain make it a suitable technology to facilitate the circular economy. Blockchain facilitates the monitoring of items throughout their lifecycle by establishing an immutable record of product details and supply chain operations. Traceability guarantees that resources may be efficiently recycled and reused, as stakeholders have access to a detailed record of a product's path, including its source, composition, and prior applications (Falco, 2023).

Blockchain technology not only facilitates traceability but also enhances transparency in the circular economy. The decentralized and transparent characteristics of blockchain enable complete visibility across the whole product lifecycle, ensuring that the origin and sustainability claims of products are publicly accessible and can be verified. The enhanced level of transparency fosters trust among stakeholders, especially customers who have a growing need for detailed information regarding the origins and life cycle of the items they buy (Wognum et al., 2011).

In addition, blockchain can serve as a reliable database, where smart contracts can automate the transfer of information while upholding data privacy considerations. The predetermined regulations for accessing and exchanging data can be implemented via the blockchain, enabling streamlined and reliable cooperation among parties in the supply chain. Trusted data sharing is essential for the circular economy since it allows for the smooth exchange of information and the coordination of circular activities (Rejeb et al., 2022).

By integrating traceability solutions based on blockchain technology, companies can enhance their reputation and accountability within the circular economy. The transparent and verifiable history of sustainable actions that blockchain technology offers can enhance confidence in a company's dedication to circular principles, establishing a more solid basis for cooperation and the implementation of sustainable practices (Juszczyk & Shahzad, 2022).

Smart contracts have become a transformative method for improving efficiency, transparency, and accountability in resource management within the circular economy. Smart contracts, which utilize blockchain technology, provide the ability to create agreements that are capable of executing themselves based on predetermined rules and norms. The automated execution is crucial in optimizing the allocation of resources, especially in the context of IoT devices and supply chain processes (Majeed & Rupasinghe, 2017).

Smart contracts are utilized inside the circular economy framework to enforce regulations and policies that regulate the behavior of IoT devices. This ensures that scheduling rules and resource management protocols are followed securely and automatically. Through the implementation of smart contracts, stakeholders can create a system that ensures indisputable and clear execution of resource allocation processes. This enables effective monitoring of resource usage and the development of credit-based resource management systems that are connected to blockchain technology (Pincheira et al., 2021).

Furthermore, the use of deserving resource smart contracts (DRSC) and punishment smart contracts (PSC) in supply chain resource management systems demonstrates the capacity of smart contracts to transform resource allocation methods. These contracts are specifically created to efficiently oversee transactions, guaranteeing equitable and effective allocation of resources according to predetermined criteria. By utilizing smart contracts, organizations can optimize resource allocation procedures, strengthen accountability, and enhance overall efficiency in resource management systems (Kaya, 2021).

Tokenization of assets is the act of generating digital tokens on a blockchain to symbolize ownership rights or assets, regardless of whether they are physical or digital in nature. This novel strategy utilizes smart contract and blockchain technology to create tokens for assets, allowing for the representation of ownership or rights to a token on the blockchain. Asset tokenization refers to the process of converting various types of assets into digital tokens. These assets might include physical assets like real estate, commodities, and art, as well as financial assets like stocks and bonds (Kuchanur, 2015). Additionally, non-physical assets like digital art and intellectual property can also be tokenized. Tokenizing assets involves representing ownership rights as digital tokens kept on a blockchain, which enables enhanced liquidity, transparency, and efficiency in managing and conducting transactions involving assets (Tian et al., 2020).

7.5   POTENTIAL BENEFITS AND OPPORTUNITIES

Blockchain technology offers numerous advantages in facilitating the shift toward a circular economy (Figure 7.2). One of the main benefits is the possibility of less counterfeiting and fraud. The transparent and irreversible ledger of blockchain technology can effectively address the issue of counterfeiting by offering a secure and unchangeable record of a product's origin and genuineness. Ensuring the integrity of recycled and reused materials is vital in the circular economy, as the traceability of resources is key for efficient recycling and reuse (Falco, 2023).

Blockchain technology in the circular economy also enhances resource efficiency. Blockchain technology has the potential to improve transparency and traceability in supply chains, allowing stakeholders to more effectively monitor the movement of materials and products. Enhancing visibility can effectively enhance the utilization of resources, minimize inefficiencies, and promote the reutilization and recycling of materials, so fostering a more sustainable and circular economy (Awan & Sroufe, 2022).

Moreover, blockchain technology has the potential to increase consumer involvement by enabling customers to make better-informed purchasing choices. Blockchain technology enhances transparency and empowers consumers to make informed choices on sustainable and circular products by offering comprehensive details about the product's origin, materials, and sustainability credentials. This can motivate enterprises to embrace circular processes and facilitate the shift toward a more sustainable economic model (Boukis, 2020).

By harnessing the fundamental characteristics of blockchain, such as its decentralized nature, transparent operation, and immutable record-keeping, businesses and policymakers may tap into fresh possibilities to promote a circular and sustainable economy. Ultimately, this will aid in the worldwide endeavor to achieve a more sustainable future (Montanari, 2022).

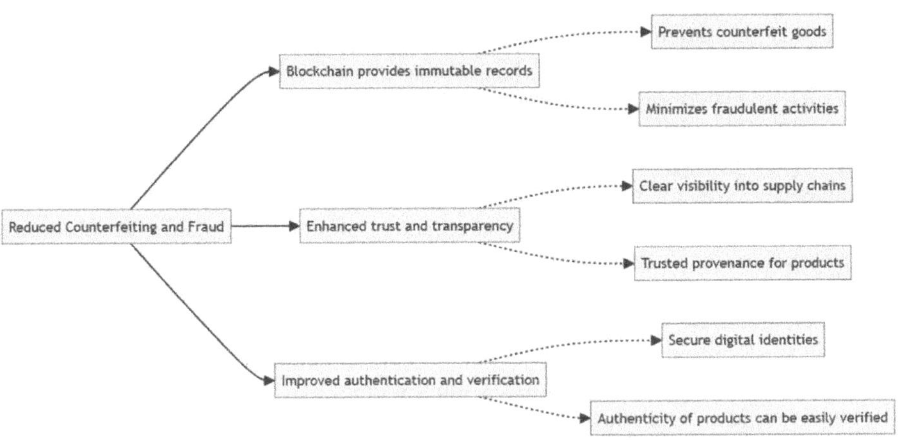

FIGURE 7.2   Potential advantages and possibilities.

7.6   CURRENT PROJECTS AND INNOVATIONS

Blockchain technology has become a powerful tool for improving supply chain management by providing unparalleled transparency, trust, and efficiency across the whole supply chain process. Blockchain-based traceability solutions are transforming the process of tracking the movement of products, commodities, and transactions for businesses and consumers (Chang et al., 2019).

A solution called wholechain is available, which is a platform based on blockchain technology that simplifies the process of creating digital replicas of things. Wholechain utilizes QR code scanning and blockchain technology to swiftly access comprehensive product records, facilitating streamlined supply chain mapping and heightened transparency. The blockchain's decentralized and irreversible characteristics guarantee that all transactions and product information are securely documented, promoting trust and responsibility among supply chain collaborators (Melo et al., 2019).

IBM blockchain is another notable example that provides a range of supply chain solutions that utilize the capabilities of smart contracts. These autonomous, self-executing contracts, which are established on predetermined circumstances, provide almost immediate insight into supply chain activities, enabling stakeholders to make proactive decisions in the event of deviations. IBM's blockchain solutions facilitate the secure exchange of reliable information, thereby fostering responsible sourcing, ensuring product authenticity, and reducing conflicts.

The utilization of cutting-edge blockchain technology is revolutionizing conventional methods of managing supply chains. These solutions offer a more streamlined, secure, and transparent approach to monitoring items and transactions at every stage of the supply chain process. As organizations and sectors increasingly adopt blockchain technology, the incorporation of blockchain-based supply chain tracking is becoming an essential element in the shift toward a more sustainable and circular economy (Grida & Mostafa, 2023).

Waste Management and Recycling Initiatives involve a wide range of methods and actions that attempt to promote environmental conservation and sustainable waste management. These projects prioritize trash reduction, improve recycling endeavors, and encourage acceptable waste disposal methods to limit environmental harm and build a circular economy. An essential element of these projects entails the segregation of waste at its source to avoid the pollution of recycling channels and guarantee the effectiveness of recycling procedures. Efficient waste separation is crucial for optimizing the retrieval and recycling of materials, making a substantial contribution to sustainable waste management methods (Zorpas, 2020).

Efforts to minimize the utilization of paper and plastic products are fundamental elements of waste management strategies. Organizations and people can decrease trash output and reduce their environmental impact by advocating for the reduction of paper use and supporting the use of eco-friendly alternatives, such as digital documents and reusable items. Similarly, efforts focused on diminishing plastic consumption and promoting alternatives to disposable plastics are crucial in waste management methods. By adopting sustainable alternatives and behaviors, such as utilizing reusable containers and compostable products, it is possible to reduce the detrimental effects of plastic waste on the environment (Nasrollahi et al., 2020).

Tackling the problem of styrofoam usage is another crucial element of waste management programs. By refraining from using styrofoam, a non-biodegradable substance that poses harm to the environment, and instead choosing recyclable paper products or other eco-friendly alternatives, one can effectively minimize waste and foster environmental sustainability. In addition, implementing initiatives such as Bring Your Own Container (BYOC) promotes the usage of reusable containers and bags, thus decreasing dependence on disposable packaging and supporting waste reduction endeavors. By integrating Bring Your Own Container (BYOC) practices into everyday activities like grocery shopping, individuals are empowered to make environmentally conscious decisions and reduce the amount of garbage they produce (Andreas et al., 2012).

The Waste Management and Recycling Initiatives emphasize the significance of both individual and collective efforts in advocating for sustainable waste practices, minimizing environmental harm, and cultivating a more circular and environmentally aware approach to trash management. By adopting these initiatives and integrating sustainable practices into their everyday routines, individuals and organizations may make a valuable contribution to promoting a healthier environment and ensuring a more sustainable future (Alshuwaikhat & Abubakar, 2008).

Blockchain technology enables circular business models that utilize its unique characteristics to promote sustainability and circularity across different industries. Businesses can use new strategies to enhance resource efficiency, minimize waste, and encourage responsible consumption by leveraging the decentralized and transparent ledger system of blockchain. These circular business models frequently entail the process of tokenizing products, which allows for traceability and promotes collaboration among stakeholders, ultimately leading to the establishment of a more sustainable and environmentally aware ecosystem. Blockchain enables the verification of product origin, environmental impact, and ownership across the whole lifespan of a product. This allows firms to implement circular practices including product reuse, recycling, and material recovery. By incorporating blockchain technology, circular business models gain the ability to improve supply chain transparency, maximize resource utilization, and facilitate the shift toward a more sustainable and circular economy (Abideen et al., 2021).

7.7 FUTURE OUTLOOK AND CHALLENGES

The future prospects and obstacles related to scalability concerns in blockchain technology are crucial factors to consider for the ongoing advancement and acceptance of blockchain-based solutions. As the user base of blockchain networks increases, the issue of scalability becomes a prominent challenge that needs to be resolved in order to effectively and extensively implement blockchain technology in many industries. The scalability issue in blockchain is complex, involving various elements such as cost, capacity, networking, and throughput limits that affect the network's capacity to efficiently manage growing transaction volumes (Sanka & Cheung, 2021).

The issue of scalability is especially apparent in the realm of blockchain networks such as Bitcoin and Ethereum, where the volume of daily transactions has significantly increased over time, emphasizing the necessity for scalable solutions to support this expansion. The growing use of blockchain systems, together with the increase in daily transactions, highlights the need to create scalable blockchain designs that can meet the changing needs of users and applications. As blockchain technology progresses and extends beyond its original emphasis on the financial sector, the ability to scale becomes a crucial obstacle that needs to be solved in order to fully utilize the promise of blockchain in many industries (Al-Jaroodi & Mohamed, 2019).

To tackle scalability challenges, it is necessary to develop creative solutions and make progress in blockchain design to better network performance, decrease transaction costs, and enhance transaction completion times. Efforts are underway to address the scalability issue in the blockchain ecosystem by investigating and adopting measures like sharding, lightning networks, and other scaling solutions. These initiatives aim to create a more sustainable, efficient, and secure blockchain environment. The blockchain community is prioritizing scalability improvements to establish a strong basis for the mainstream use of blockchain technology. This will facilitate smooth interactions, safe transactions, and improved scalability across various applications and industries (Bhushan et al., 2020).

The regulatory and legal framework around blockchain technology is an intricate and dynamic field that necessitates meticulous maneuvering to guarantee adherence to rules and promote creativity. With the increasing prevalence of blockchain applications in several industries, it is crucial for enterprises, governments, and stakeholders to have a comprehensive understanding of the regulatory and legal aspects involved (Toufaily et al., 2021).

Data privacy and protection are a crucial aspect of regulatory discussions. Given the unchangeable and easily visible characteristics of blockchain, there are worries about how personal data are managed and if it complies with data protection legislation. Current initiatives are being pursued to investigate privacy-enhancing technologies and methods that can effectively balance openness and data privacy rights. The aim is to ensure that blockchain applications comply with legal obligations while preserving the integrity of the technology (Yeoh, 2017).

The blockchain industry faces notable difficulties regarding intellectual property rights, namely related to safeguarding patents, preventing copyright violations, and determining ownership of digital assets. Regulatory frameworks are being created to tackle these problems and establish clear guidelines on intellectual property rights in blockchain networks, protecting innovation and promoting a favorable climate for technical progress (Gürkaynak et al., 2018).

In addition, the regulatory environment for digital currencies, smart contracts, and cybersecurity is changing quickly as governments and regulatory agencies strive to create explicit norms and frameworks. Cryptocurrency rules often encompass anti-money laundering protocols, securities legislation, and investor protection procedures to limit risks and uphold market integrity. Efforts are underway to thoroughly examine the legal aspects concerning smart contracts, including contractual intent and culpability in the event of disputes. The aim is to establish legal clarity and promote the use of blockchain-based contracts (Drummer & Neumann, 2020).

Effective collaboration among stakeholders, policymakers, regulators, and legal experts is crucial in order to establish a regulatory framework that effectively balances innovation and compliance. This collaboration is necessary to promote responsible adoption of blockchain technology and to solve the legal difficulties that arise from its revolutionary nature. By strategically and preemptively managing the legislative and legal framework, the blockchain ecosystem may flourish, fostering sustainable expansion and ingenuity across several industries (Refaei, 2023).

The extensive implementation of blockchain technology encounters various substantial obstacles, as companies and organizations struggle with the difficulties of embracing this revolutionary breakthrough. A major obstacle to adoption is the innate reluctance to change that frequently exists in well-established sectors and organizational cultures (Heidenreich & Talke, 2020).

Professionals and stakeholders who are familiar with conventional processes and workflows may be reluctant to adopt the transformative potential of blockchain technology. The opposition observed could be attributed to concerns about potential job displacement or the requirement to adjust to novel work methodologies, along with a general lack of comprehension of the potential advantages and consequences of blockchain technology in their particular sectors (Toufaily et al., 2021).

Moreover, the perceived intricacy and technicality of blockchain might lead to a feeling of confusion and dread among business participants. The apprehension toward unfamiliarity, along with insufficient and ambiguous communication and instruction, can impede the seamless assimilation of blockchain solutions into current systems and organizational frameworks (Naiseh et al., 2024).

To overcome these obstacles in adoption, a comprehensive strategy is needed that tackles the underlying reasons for reluctance. To create a culture that is open to technological change and innovation, it is essential to have effective communication, thorough training programs, and active involvement of stakeholders in the implementation process (Myeong et al., 2021).

To facilitate the implementation of blockchain technology, firms should address issues related to job security, workflow disruptions, and the broader implications of this transformational invention. In addition, emphasizing the concrete advantages of blockchain, such as enhanced productivity, visibility, and financial savings, might mitigate industry opposition and promote the widespread use of this innovative technology (Li & Kassem, 2021).

In order to fully harness the promise of blockchain technology and successfully integrate it into various industries, it is crucial to overcome the challenges of acceptance and industry resistance that currently exist within the evolving blockchain ecosystem. By actively and preemptively tackling these obstacles, firms may establish themselves as leaders in technological innovation and take advantage of the game-changing prospects offered by blockchain technology (Kumar et al., 2024).

7.8 SUMMARY

Incorporating blockchain technology into the circular economy offers a significant chance to alter conventional linear models and advance sustainable practices. The intrinsic characteristics of blockchain, such as transparency and traceability, perfectly match the fundamental principles of the circular economy. This allows stakeholders to easily monitor the progress of products and materials at every step of their lifecycle.

To summarize the main aspects emphasized in the sources, blockchain technology has the ability to greatly support the transition to a circular economy by promoting cleaner economic transactions and developing a harmonious relationship between the environment, economy, and society. The cooperative effort needed to utilize the overall advantages of blockchain highlights the significance of group endeavors in promoting innovation and sustainability inside the circular economy framework.

In order to advance blockchain technology in the circular economy, it is crucial to maintain a strong emphasis on collaboration between stakeholders, policymakers, industry professionals, and researchers. Through collaborative efforts to overcome barriers to adoption, industry opposition, regulatory obstacles, and scalability concerns, the blockchain ecosystem has the potential to establish a more environmentally friendly, streamlined, and accountable circular economy.

By adopting collaboration as a fundamental principle, the potential of blockchain in the circular economy is vast. It has the ability to bring about significant positive transformation, enhance resource efficiency, and establish an economic model that is both environmentally sensitive and socially responsible. By collaborating and prioritizing creative solutions, blockchain technology may have a significant impact on promoting the circular economy and defining a sustainable future for future generations.

REFERENCES

Abideen, A. Z., Pyeman, J., Sundram, V. P. K., Tseng, M.-L., & Sorooshian, S. (2021). Leveraging capabilities of technology into a circular supply chain to build circular business models: A state-of-the-art systematic review. *Sustainability*, *13*(16), 8997. www.mdpi.com/2071-1050/13/16/8997

Al-Jaroodi, J., & Mohamed, N. (2019). Blockchain in industries: A survey. *IEEE Access*, *7*, 36500–36515. https://doi.org/10.1109/ACCESS.2019.2903554

Alshuwaikhat, H. M., & Abubakar, I. (2008). An integrated approach to achieving campus sustainability: Assessment of the current campus environmental management practices. *Journal of Cleaner Production*, *16*(16), 1777–1785. https://doi.org/10.1016/j.jclepro.2007.12.002

Anbumozhi, V., & Kimura, F. (2018). Industry 4.0: What does it mean for the circular economy in ASEAN? *Industry 4.0: Empowering ASEAN for the Circular Economy*, 1.

Andreas, A., Gumulia, I., Lunn, D., & Nguyen, T. (2012). *An investigation into the Bring Your Own Container project implementation* [Text]. https://open.library.ubc.ca/collections/18861/items/1.0108456

Awan, U., & Sroufe, R. (2022). Sustainability in the circular economy: Insights and dynamics of designing circular business models. *Applied Sciences*, *12*(3), 1521.

Bhatt, T., Cusack, C., Dent, B., Gooch, M., Jones, D., Newsome, R., . . ., Zhang, J. (2016). Project to develop an interoperable seafood traceability technology architecture: Issues brief. *Comprehensive Reviews in Food Science and Food Safety*, *15*(2), 392–429.

Bhushan, B., Khamparia, A., Sagayam, K. M., Sharma, S. K., Ahad, M. A., & Debnath, N. C. (2020). Blockchain for smart cities: A review of architectures, integration trends and future research directions. *Sustainable Cities and Society*, *61*, 102360. https://doi.org/10.1016/j.scs.2020.102360

Boukis, A. (2020). Exploring the implications of blockchain technology for brand – consumer relationships: A future research agenda. *Journal of Product & Brand Management*, *29*(3), 307–320.

Bressanelli, G., Perona, M., & Saccani, N. (2019). Challenges in supply chain redesign for the circular economy: A literature review and a multiple case study. *International Journal of Production Research*, *57*(23), 7395–7422.

Bucknall, D. G. (2020). Plastics as a materials system in a circular economy. *Philosophical Transactions of the Royal Society A*, *378*(2176), 20190268.

Centobelli, P., Cerchione, R., Del Vecchio, P., Oropallo, E., & Secundo, G. (2022). Blockchain technology for bridging trust, traceability and transparency in circular supply chain. *Information & Management*, *59*(7), 103508.

Chang, S. E., Chen, Y.-C., & Lu, M.-F. (2019). Supply chain re-engineering using blockchain technology: A case of smart contract based tracking process. *Technological Forecasting and Social Change, 144*, 1–11. https://doi.org/10.1016/j.techfore.2019.03.015

Cobîrzan, N., Muntean, R., & Felseghi, R.-A. (2023). *Circular economy implementation for sustainability in the built environment.* IGI Global.

Dapp, M. M. (2018). Toward a sustainable circular economy powered by community-based incentive systems. In *Business transformation through blockchain* (Vol. 2, pp. 153–181). Springer.

de Sa, P., & Korinek, J. (2021). Resource efficiency, the circular economy, sustainable materials management and trade in metals and minerals. In *OECD trade policy papers, No. 245.* OECD Publishing. https://doi.org/10.1787/69abc1bd-en

Drummer, D., & Neumann, D. (2020). Is code law? Current legal and technical adoption issues and remedies for blockchain-enabled smart contracts. *Journal of Information Technology, 35*(4), 337–360. https://doi.org/10.1177/0268396220924669

Dsouza, A., Price, G. W., Dixon, M., & Graham, T. (2021). A conceptual framework for incorporation of composting in closed-loop urban controlled environment agriculture. *Sustainability, 13*(5), 2471.

Falco, F. (2023). *Distributed ledger technology in the circular economy: Enable traceability and transparency of recyclable products with a digital product passport platform.* FH Vorarlberg (Fachhochschule Vorarlberg).

Fan, E., Li, L., Wang, Z., Lin, J., Huang, Y., Yao, Y., . . ., Wu, F. (2020). Sustainable recycling technology for Li-ion batteries and beyond: Challenges and future prospects. *Chemical Reviews, 120*(14), 7020–7063.

Feng, H., Wang, X., Duan, Y., Zhang, J., & Zhang, X. (2020). Applying blockchain technology to improve agri-food traceability: A review of development methods, benefits and challenges. *Journal of Cleaner Production, 260*, 121031.

Garcia-Torres, S., Albareda, L., Rey-Garcia, M., & Seuring, S. (2019). Traceability for sustainability – literature review and conceptual framework. *Supply Chain Management: An International Journal, 24*(1), 85–106.

Grida, M., & Mostafa, N. A. (2023). Are smart contracts too smart for supply chain 4.0? A blockchain framework to mitigate challenges. *Journal of Manufacturing Technology Management, 34*(4), 644–665. https://doi.org/10.1108/JMTM-09-2021-0359

Gürkaynak, G., Yılmaz, İ., Yeşilaltay, B., & Bengi, B. (2018). Intellectual property law and practice in the blockchain realm. *Computer Law & Security Review, 34*(4), 847–862. https://doi.org/10.1016/j.clsr.2018.05.027

Heidenreich, S., & Talke, K. (2020). Consequences of mandated usage of innovations in organizations: Developing an innovation decision model of symbolic and forced adoption. *AMS Review, 10*(3), 279–298. https://doi.org/10.1007/s13162-020-00164-x

Iacovidou, E., Hahladakis, J. N., & Purnell, P. (2021). A systems thinking approach to understanding the challenges of achieving the circular economy. *Environmental Science and Pollution Research, 28*, 24785–24806.

Juszczyk, O., & Shahzad, K. (2022). Blockchain technology for renewable energy: Principles, applications and prospects. *Energies, 15*(13), 4603.

Kara, S., Hauschild, M., Sutherland, J., & McAloone, T. (2022). Closed-loop systems to circular economy: A pathway to environmental sustainability? *CIRP Annals, 71*(2), 505–528.

Kaya, F. (2021). *Investigation of legal issues and contractual considerations associated with BIM use in construction projects.* Middle East Technical University.

Kuchanur, D. A. B. (2015). Analysis of investment in financial and physical assets: A comparative study. *International Journal of Research in Management, Economics & Commerce, 5*(4), 61–76.

Kumar, V., Ashraf, A. R., & Nadeem, W. (2024). AI-powered marketing: What, where, and how? *International Journal of Information Management*, 102783. https://doi.org/10.1016/j.ijinfomgt.2024.102783

Lazarevic, D., & Brandão, M. (2020). The circular economy: A strategy to reconcile economic and environmental objectives? In *Handbook of the circular economy* (pp. 8–27). Edward Elgar Publishing.

Leng, J., Ruan, G., Jiang, P., Xu, K., Liu, Q., Zhou, X., & Liu, C. (2020). Blockchain-empowered sustainable manufacturing and product lifecycle management in Industry 4.0: A survey. *Renewable and Sustainable Energy Reviews*, *132*, 110112.

Li, J., & Kassem, M. (2021). Applications of distributed ledger technology (DLT) and blockchain-enabled smart contracts in construction. *Automation in Construction*, *132*, 103955. https://doi.org/10.1016/j.autcon.2021.103955

Liu, X., Lu, Y., Iseri, E., Shi, Y., & Kuzum, D. (2018). A compact closed-loop optogenetics system based on artifact-free transparent graphene electrodes. *Frontiers in Neuroscience*, *12*, 322176.

Madaan, G., Singh, A., Mittal, A., & Shahare, P. (2024). Reduce, reuse, recycle: Circular economic principles, sustainability and entrepreneurship in developing ecosystems. *Journal of Small Business and Enterprise Development*, *31*(6), 1041–1066. https://doi.org/10.1108/JSBED-01-2023-0009

Majeed, A. A., & Rupasinghe, T. D. (2017). Internet of things (IoT) embedded future supply chains for Industry 4.0: An assessment from an ERP-based fashion apparel and footwear industry. *International Journal of Supply Chain Management*, *6*(1), 25–40.

Melo, W. S., Bessani, A., Neves, N., Santin, A. O., & Carmo, L. F. R. C. (2019). Using blockchains to implement distributed measuring systems. *IEEE Transactions on Instrumentation and Measurement*, *68*(5), 1503–1514. https://doi.org/10.1109/TIM.2019.2898013

Mestre, A., & Cooper, T. (2017). Circular product design: A multiple loops life cycle design approach for the circular economy. *The Design Journal*, *20*(sup 1), S1620–S1635.

Montanari, P. (2022). The impact of blockchain and DLT enterprise tokens on business processes. *Analysis of applications of different use cases in major industries*.

Myeong, S., Kim, Y., & Ahn, M. J. (2021). Smart city strategies – technology push or culture pull? A case study exploration of Gimpo and Namyangju, South Korea. *Smart Cities*, *4*(1), 41–53. www.mdpi.com/2624-6511/4/1/3

Naiseh, M., Clark, J., Akarsu, T., Hanoch, Y., Brito, M., Wald, M., . . . Shukla, P. (2024). Trust, risk perception, and intention to use autonomous vehicles: An interdisciplinary bibliometric review. *AI & Society*. https://doi.org/10.1007/s00146-024-01895-2

Nasrollahi, M., Beynaghi, A., Mohamady, F. M., & Mozafari, M. (2020). Plastic packaging, recycling, and sustainable development. In W. Leal Filho, A. M. Azul, L. Brandli, P. G. Özuyar, & T. Wall (Eds.), *Responsible consumption and production* (pp. 544–551). Springer International Publishing. https://doi.org/10.1007/978-3-319-95726-5_110

Nogueira, A., Ashton, W., Teixeira, C., Lyon, E., & Pereira, J. (2020). Infrastructuring the circular economy. *Energies*, *13*(7), 1805.

Park, M. B. (2009). *Product life: Designing for longer lifespans*. Kingston University.

Pincheira, M., Vecchio, M., Giaffreda, R., & Kanhere, S. S. (2021). Cost-effective IoT devices as trustworthy data sources for a blockchain-based water management system in precision agriculture. *Computers and Electronics in Agriculture*, *180*, 105889.

Refaei, D. M. D. M. (2023). Regulatory frameworks for autonomous robotics in NEOM's sustainable technology landscape. *Migration Letters*, *20*(9), 228–258. https://doi.org/10.59670/ml.v20i9.5965

Rejeb, A., Zailani, S., Rejeb, K., Treiblmaier, H., & Keogh, J. G. (2022). Modeling enablers for blockchain adoption in the circular economy. *Sustainable Futures*, *4*, 100095.

Sanka, A. I., & Cheung, R. C. C. (2021). A systematic review of blockchain scalability: Issues, solutions, analysis and future research. *Journal of Network and Computer Applications, 195*, 103232. https://doi.org/10.1016/j.jnca.2021.103232

Tian, Y., Adriaens, P., Minchin, R. E., Chang, C., Lu, Z., & Qi, C. (2020). Asset tokenization: A blockchain solution to financing infrastructure in emerging markets and developing economies. *ADB-IGF special working paper series "fintech to enable development, investment, financial inclusion, and sustainability"*.

Toufaily, E., Zalan, T., & Dhaou, S. B. (2021). A framework of blockchain technology adoption: An investigation of challenges and expected value. *Information & Management, 58*(3), 103444. https://doi.org/10.1016/j.im.2021.103444

Van Ewijk, S., & Stegemann, J. (2016). Limitations of the waste hierarchy for achieving absolute reductions in material throughput. *Journal of Cleaner Production, 132*, 122–128.

Wognum, P. N., Bremmers, H., Trienekens, J. H., Van Der Vorst, J. G., & Bloemhof, J. M. (2011). Systems for sustainability and transparency of food supply chains – current status and challenges. *Advanced Engineering Informatics, 25*(1), 65–76.

Xiong, X., Liu, X., Iris, K., Wang, L., Zhou, J., Sun, X., . . . Lin, Z. (2019). Potentially toxic elements in solid waste streams: Fate and management approaches. *Environmental Pollution, 253*, 680–707.

Xu, Y., Zhang, C., Zeng, Q., Wang, G., Ren, J., & Zhang, Y. (2020). Blockchain-enabled accountability mechanism against information leakage in vertical industry services. *IEEE Transactions on Network Science and Engineering, 8*(2), 1202–1213.

Yeoh, P. (2017). Regulatory issues in blockchain technology. *Journal of Financial Regulation and Compliance, 25*(2), 196–208. https://doi.org/10.1108/JFRC-08-2016-0068

Yildizbasi, A. (2021). Blockchain and renewable energy: Integration challenges in circular economy era. *Renewable Energy, 176*, 183–197.

Zorpas, A. A. (2020). Strategy development in the framework of waste management. *Science of the Total Environment, 716*, 137088. https://doi.org/10.1016/j.scitotenv.2020.137088

8 Sustainability Reporting and Transparency

*Seyed Yasin Jamali, Aydin Shishegaran,
Jack Smith, and Mohsen Saeedi*

8.1 INTRODUCTION

In the current environment of growing environmental concern, social and economic issues, and development with respect to sustainability, especially the development of corporate governance processes, sustainability has become one of the most critical and dynamic concepts influencing both business and society (Hahn and Kühnen, 2013). Correspondingly, sustainability reporting stands for a well-structured report through which an organization discloses its environmental, social, and governance (ESG) performance (Shen and Pena-Mora, 2018). This concept began to develop in the last decades of the 20th century, driven by increasing societal expectations, regulatory pressures, and the recognition of sustainability as a strategic business imperative (Schaltegger et al., 2016).

On the one hand, sustainability reports are policy reflections of the governments on sustainability development. This policy results in the possibility of sustainable evaluation for various groups and other stakeholders' possibility to evaluate a company's impact on the environmental, social, and economic situations for the next generations with respect to this view that there is only a planet to live and prepare our requirements and the requirements of the next generations, but its resource is not unlimited (Shishegaran et al., 2022; Nidumolu et al., 2009).

Although sustainability reporting's history is relatively brief, over the past few decades, it has developed from being on the fringes to being in the forefront (Adams and Frost, 2008). There are several reasons for its emergence within the broader societal context, including reshaping societal expectations, regulatory constraints, the growing semantics of sustainability, and business strategy's meaning (Hahn and Figge, 2011). These processes led to the appearance of numerous frameworks and guidelines specifically designed to mitigate sustainability reporting's drawbacks, enhance efficiency, and improve the quality bar overall.

Transparency is a crucial success factor in sustainability reporting and, at the same time, a challenge for it (Dillard and Vinnari, 2017). Transparency within the field of reporting is when relevant information is faithfully and precisely disclosed on time. It builds the trust of the broader public and empowers stakeholders to form wise judgments about organizations' deeds. Making sustainability reporting transparent implies overcoming numerous obstacles that remain a hindrance. These are scattered information, organizations' fears about the negative impact of non-sham

DOI: 10.1201/9781003609865-8

report disclosures, resource-central reporting tendencies, and more (Adams and Frost, 2008).

Blockchain is a recent technology and, at the same, an impactful novelty due to its transparency-enhancing potential. It is considered relevant to many other writers and professionals because the unique features of blockchain technology are precisely advocating those of transparency: trackability and immutability (Li and Badinelli, 2022). This chapter represents a comprehensive knowledge of the current state of the art regarding blockchain-promoted transparency in the reporting of sustainability. The authors' objective is precisely to aggregate the most crucial studies, theoretical frames, and case studies on this subject to identify the relations between sustainability reporting and blockchain and how both can help improve environmental, social, and economic issues in the development. Also, this chapter clarifies the concepts of sustainability reporting and blockchain and their effect on our world.

8.2   SUSTAINABILITY

Sustainability reporting is defined as the systematic disclosure of an organization's ESG business practices (Schaltegger and Burritt, 2006). It has traditionally been seen as something that only a few companies go to great lengths to do, but that has changed in the last few years to become required of businesses in most countries across the globe (KPMG International, 2020).

In the 1960s and 1970s, with the start of environmental movements, public awareness of environmental degradation and limited natural resources increased. The publications such as Rachel Carson's "Silent Spring", Robert L. Rudd's "Pesticides and the Living Landscape", and Paul Ehrlich's "The Population Bomb" helped this matter a lot. One of the best definitions for sustainable development is the definition of the World Commission on Environment and Development, which was in the Brundtland Report in 1987: "the needs of the present without compromising the ability of future generations to meet their own needs" (World Commission on Environment and Development, 1987).

8.2.1   KEY CONCEPTS IN SUSTAINABILITY

Early work on sustainability reporting concentrated on formulating frameworks and rules to ensure consistency and comparability across businesses (Lozano and Huisingh, 2011). Two technologies that put the squeeze on the sustainability reporting space early were the Global Reporting Initiative (GRI) and the United Nations Sustainable Development Goals (SDGs). GRI was established in 1997 and is a collection of guidelines or standards developed to help organizations record, evaluate, and disclose their sustainability achievements. The standards encompass multiple topics including, but not limited to, environmental impacts (such as energy use, greenhouse gas emission, and waste management), social impacts (such as labor practices, human rights, and engagement with communities), and governance issues ranging from management, ethics, and anti-corruption, to board composition. GRI helps companies communicate their sustainability metrics enough clearly, comprehensively, and comparably with this standardized reporting system (Global Reporting

Initiative (GRI), 2016). However, in addition to GRI, hundreds of other standards and frameworks focused on sustainability reporting are now available. For example, SASB sets industry-specific standards on how to report financial material sustainability information. CDP collects environmental data on behalf of institutional investors regarding global climate change, water use, and security as well as deforestation. IIRC advocates and advances integrated reporting on an organization's strategy, governance, performance, and prospects. TCFD legalizes the required scope of disclosures that financial institutions must make regarding long-term strategy. ISO 26000 guides how organizations can integrate social responsibility into their actions. Corporate sustainability frameworks and standards such as Dow Jones Sustainability Indices, Corporate Sustainability Assessment, and Corporate Social Responsibility Indicators push companies to evaluate themselves according to their economic, environmental, and social performance (Fathi, 2021). There are several frameworks and standards on sustainability reporting that suggest to organizations a possibility to choose the most suited for them to meet the expectations and needs of a wide variety of interested stakeholders.

The set of 17 interconnected goals has been adopted by all parties to the United Nations in 2015 as a universal call to action to address poverty, inequality, climate change, environmental degradation, peace, and justice, and many companies have engaged in aligning their sustainability activities with the SDGs. Therefore, by following and using these protocols and frameworks among many others, businesses can improve their sustainability practices and inspire others as a means of realizing the balance between their sustainability and responsibility reporting in the world (United Nations, 2015).

8.2.2 Sustainable Practices in Various Industries

The rise of stakeholder interests has helped drive the growing demand for regulatory sustainability (Freeman, 2010). Demands are a reflection of increasing consumer, investor, employee, and community awareness of business contribution to society (Porter and Kramer, 2011). Socioeconomic issues, such as climate change, natural disasters, food security, health concerns and more, have led to increased societal awareness about the importance of integration of sustainability (Clark et al., 2015). Governments and regulatory authorities have enacted more sustainability regulations and bonuses in recent years to force and enable companies to report their own sustainability measures (Eccles et al., 2012). Under the laws, companies may be compelled to disclose particular ESG indicators or use a reporting protocol, or they may be forced to show that they meet particular sustainable requirements (Clark and Yavuz, 2018). Lawful requirements enforce stricter accountability and foster transparency.

Organizations are realizing that embedding sustainability in their core business models is not just ethical but a necessary aspect of their operations. Being environmentally conscious can boost innovation, eliminate expenses, enhance reputation, entice top-level people, and hedge hazards (Hahn et al., 2015). As a result, many businesses use sustainability reporting as a tool to communicate their commitment to social responsibility to external stakeholders and differentiate themselves from other companies.

The high-level takeaway is that GRI and SDGs accelerated the pace at which businesses had to improve. Specifically, their integrative pressure made firms significantly advance their sustainability reporting practices, strategically adjust to fit into the global sustainability matrix, and concretely commit to a greener and fairer future. The set of 17 interconnected goals has been adopted by all parties to the United Nations in 2015 as a universal call to action to address poverty, inequality, climate change, environmental degradation, peace, and justice, and many companies have engaged in aligning their sustainability activities with the SDGs. Therefore, by following and using these protocols and frameworks among many others, businesses can improve their sustainability practices and inspire others as a means of realizing the balance between their sustainability and responsibility reporting in the world (United Nations, 2015).

8.2.3 THE COMPLEX LANDSCAPE OF SUSTAINABILITY REPORTING

From the point of view of sustainability reporting, it should be noted that developing a sustainable product is very complicated as there are many parts in every supply chain that have their own environmental impacts and thus appears to be hard to control the chains. Supply chains present a high level of complexity and integration that might involve multiple countries and comprise several other firms such as suppliers, manufacturers, distributors, or retailers. All these firms take part in the overall environmental footprint of the final good, as they all use it in their work processes. The production, including extraction of raw materials, manufacturing process, transportation, packaging materials, and waste disposal, all are responsible for the outcomes, which might be harmful to the environment. For example, manufacturing is generally connected to waste, pollution, and greenhouse gas emissions. Transportation operates on carbon and energy, and finally, packaging material extraction and production are also related to waste and pollution. Emission reduction is crucial for sustainability (IPCC, 2014; Cardu et al., 2023; Paris Agreement, 2015), and it is vital for any organization to effectively assess and address their footprint, and try to minimize the environmental harms (UNEP, 2019). The scope one emissions are all sources of emission under the control of the part of the production or consumption in question. This includes fuel combustion in company vehicles, combustion from company onsite energy generation, and all industrial processes delivered by the organization. The scope two emissions are all of the other company's indirect emissions due to purchased electricity, heat, or other forms of energy, which despite occurring from other facilities, it forms a relationship with the company's sustainability efforts, and finally the scope three emissions are associated with the inputs which are not directly owned or controlled by the organization, like emissions from transportation, supply chain, etc. (World Resources Institute & World Business Council for Sustainable Development, 2011).

It is difficult to perform the evaluation of scopes 2 and 3 since the emissions from them are made indirectly (Hertwich and Wood, 2018; DP, 2019). Energy providers in different countries follow different standards, which will lead to information dispersion. On the other hand, ensuring the correctness and correctness of the collected information is a tedious and difficult task due to its huge scope, which leads to more manpower and additional costs, and sometimes it can even multiply the costs of the company. Overall, the cost of collecting, managing, reporting data, third-party verification, stakeholder engagement processes, and extensive reports are significant (Naheed et al., 2024), which can lead to the loss of the company's motivation to allocate more resources and causing this scope to be neglected.

When it comes to the many stakeholder relationships, one can name the complexity of sustainability goals. Many stakeholders, or parties involved in the supply chain, may not see eye to eye on respective interests and priorities. It is especially the case regarding economic incentives and those, striving for environmental and social sustainability. Even when companies try to enforce more eco-friendly policies, suppliers or subcontractors can push back agreeing to these conditions as it may be costly or require significant resource allocation (Manetti, 2011).

8.3 TRANSPARENCY IN SUSTAINABILITY REPORTING

Sustainability reporting transparency is the potential of the firm to disclose information encompassed in its environmental, social, recycling, and governance activities, policies, execution, and impacts (Adams, 2004). In this prospect, transparency is the essential focus of clarifying and providing information that is fully clear, accurate, and wide so that stakeholders can clearly understand the firm's sustainability action, check progress, and demand the authority organization.

As a valuable companion of sustainability, transparency makes it necessary to develop trust, responsibility, and ethical business. If an organization is transparent in terms of sharing information regarding its ESG, it is ready to maintain ethical standards and ensure the stakeholders' or investors' trust (Hussain et al., 2019). It means it enables stakeholders, including investors, customers, employees, regulators, and communities, to make informed decisions and assess the long-term sustainability of organizations. In this light, every side can open a dialog with them and each other from scratch, share opinions or concerns, and find the most suitable way of addressing and solving priority issues (Freeman et al., 2010). Besides, transparency is associated with continuing sustainability, which conditions successful assessments, further development steps, and monitoring the progress. In the long-term perspective, if an organization follows all the rules of transparency in sustainability reporting, it will improve performance, evidence, status, and competitive advantage (Eccles and Serafeim, 2013). All these points are relevant to defining transparency as the primary driver in developing conscious, responsible, and sustainable behavior.

8.3.1 CHALLENGES AND INNOVATIONS

Despite the fact that transparency is one of the great characteristics of sustainability reporting, there are also considerable challenges in this direction. As a result, it is essential to identify and understand the drawbacks of transparency to attain sustainability and regain society's trust.

First of all, we can point out that transparency increases stakeholder involvement in projects, where may be various questions for the stakeholders, customers, and the public about the approaches of the company's functions, which can be resource-intensive to manage (Broeder, 2024). Additionally, disclosing some of the sustainability information might raise competition concerns. Consequently, companies avoid voluntarily disclosing sustainability information because they are deterred by competition issues (Mejbel and Salman, 2024). Another notable problem is called "greenwash" which is where firms or institutions may overplay their environmental or social responsibility to appear more greenly solicitous than reality would portend for them (de Freitas Netto et al., 2020). This in turn breeds a totally different kind of understanding and trust toward data transparency which can lead to mistrust among stakeholders. Finally, compliance-related barrier-to-entry could become an issue for smaller firms or several firms which operate in more than one jurisdiction. Those companies which avoid including sustainability information in their reports decline the opportunity to become more transparent and trustworthy (de Freitas Netto et al., 2020). Moreover, some other issues are connected with transforming the complex information collected into a clear and transparent one, as well as removing the information that is not needed. Difficulties in conveying complex issues like sustainability across language barriers may also lead to misunderstandings and misinterpretations (Kent et al., 2024).

There are multiple strategies and practices to overcome these challenges, as suggested by many researchers. For example, the impact of stakeholders' participation in preparing sustainability reports is very effective, leading to better transparency (Manetti, 2011; Manetti and Toccafondi, 2012). Since this relationship makes it possible for stakeholders to directly witness the company's commitment to transparency, it creates a mutual understanding between the company and its stakeholders and results in encouragement of stakeholders for more participation (Michelon et al., 2015). This kind of involvement of stakeholders not only helps increase the credibility of the sustainability reports but also makes the company gain a better reputation and position in the public's view (O'Dwyer et al., 2011).

Third-party verification can be another way of solving the problem to transparency of sustainability reports. Independent inspection and assurance as an external validation, can assure stakeholders that the obtained information complies with the predetermined standards by confirming the accuracy of the information (O'Dwyer et al., 2011). The presence of a neutral and unbiased authority will diminish doubts and create more transparency in the activities of sustainable projects (Manetti and Becatti, 2009).

Advanced technologies such as artificial intelligence can significantly improve the efficiency of the data collection and management process and to some extent be effective in removing the existing obstacles (International Telecommunication Union, 2020). Artificial Intelligence and big data analytics made the vast impact on the opportunity to get the better quality and more efficient sustainability reporting. However, it can only make sense in case unbiased and actual accurate information is provided, which highlights the need for transparent, trusted, and structured records.

Moreover, capacity building in corporate sustainability may lead to plain improvement and transparency in reports (Svensson and Wagner, 2021). Capacity building includes a well-structured training for employees and stakeholders for an efficient engagement in sustainability reporting. Making employees aware of sustainability reporting rules and relevant standards such as GRI, knowing the important role of sustainability in businesses, familiarizing them with and applying technology and related tools for collecting information and analyzing data, designing monitoring and evaluation systems, and development of the culture among employees and stakeholders makes companies, in addition to complying with regulations, able to observe the impact of their activities in sustainability activities and address this issue with more motivation (Patel, 2022).

It is interesting that all of the given approaches to transparency challenges can be enhanced by using blockchain. Blockchain's inherent characteristics, which are decentralization, transparency, and immutability, could be made significantly more effective if complemented with blockchain technology. To name a few, blockchain could be leveraged to transform stakeholders' participation by providing a platform where all of the stakeholders' contributions and changes are recorded and accessible to every stakeholder, leading to more trust and transparency (Smith and Lee, 2020). On the other hand, blockchain can help improve third-party verification by acting as an unbiased and corrupt-free platform (Johnson, 2019). Overall, blockchain can enhance capacity building by ensuring that training completions conducted by the organizations are genuine (Brown and Green, 2021). Generally, the blockchain is an appropriate tool for the recommended strategies in terms of making sustainability reporting more reliable and transparent.

8.4 FOUNDATIONAL ASPECTS OF BLOCKCHAIN TECHNOLOGY

Blockchain technology is a common denominator, mostly as it has previously been associated solely with Bitcoin and others. However, the technology has demonstrated that it can be utilized in a wide range of scenarios (Mougayar, 2016). Blockchain is best known for making record-keeping less secretive. Supply chains, health organizations, and the reporting system are examples of how this happens.

To begin, blockchain is a decentralized and distributed system. In other terms, it is not a central database. It is like a digital ledger that every transaction is transparently secured. In order to have a correct understanding of this platform, there is a chain. Each link of this chain contains information called a block. When a transaction happens, a new block is created, in which, in addition to the transaction information, a unique code (hash) is also placed in that block, which links the block to the previous block of the chain. In this way, each block is connected to its previous block and makes a chronological chain. When a block is added to the chain, the transaction information recorded in the block is visible to anyone but cannot be altered, unless we change the entire chain, which would require consensus from all network participants (Swan, 2015). This ensures the security of the blockchain and makes it almost impossible to tamper with. Blockchain's decentralization means that its control is not controlled by a specific entity, and the possibility of all people accessing the records in the blocks makes it transparent.

The field of the blockchain technology does not entail rules like most fields do, but there are specific fundamental principles or characteristics of how the blockchain works. One of these rules implies decentralization. It means that there are no singular points of control or failure, making the blockchain system more resilient and resistant to manipulation or censorship (Nakomoto, 2008). One of the main reasons that Bitcoin has been able to be used among all investors as a reference coin for financial transactions is its decentralization principle, which has created trust among investors. Due to the network's decentralization, no single entity or authority can control it, which makes it impossible to censor or interfere with the network's operations with the utilization of governments or any other influence (Vigna and Casey, 2015). The network of nodes, which confirms in a decentralized way using cryptographic techniques, can almost[1] never be hacked or tampered with. Bitcoin allows users to transact with one another as peers without the need to trust a central party to verify the transaction. This results from the fact that the network reaches a consensus about the state of the ledger. Thus, that removes the need to trust an agent, which gives the user financial sovereignty over their assets. A further key benefit of the decentralization feature is that the Bitcoin network can be easily accessed by anyone who has access to the internet. This was achieved as a result of Bitcoin having become a common form of digital money and payment network in regions with limited access to traditional financial services and less developed countries.

8.4.1 Blockchain's Role in Enhancing Transparency

The presented blockchain projects and many others share the same aspiration in applying blockchain technology's potential for transparency and sustainable reporting's simplification in supply chains. Blockchain technology would transform sustainability chains. The benefits would include a secure ledger of transactions, trackability throughout the general ledger and supply chain, ensuring the truth of shared data, and increased engagement with stakeholders (UNEP, 2021). Implementation of smart contracts would realize executional and compliance enforcement of sustainability agreements. It is also clear that blockchain protects its data using cryptology, reducing the number of fraud and data tampering, which would increase trust in reporting. Last but not least, a decentralized ledger reduces the cost of doing business directly because there is no documentation to verify, and this would help companies make more profit by transferring to sustainability. Every benefit solves an environmental or a social problem.

In addition to all the above, the blockchain system offers another great advantage. All consumers are able to easily monitor the manufacturing process of a product and the amount of energy consumption, from the beginning to the end. For example, when you decide to buy a piece of furniture, as the quality of construction, style, comfort and ergonomics, price and value, and after-sales service (which are usually the main factors in the choice) are easily available to you, you should be able to monitor environmental metrics easily. Blockchain allows you to go beyond different ISOs to accurately track the origin of materials, the manufacturing process, waste, transportation systems, and more.

8.4.2 Blockchain Innovations in Global Supply Chains

A few companies have begun using blockchain to do so. IBM Corporation (FPT) developed a blockchain network named IBM Food Trust to improve transparency throughout the food supply chain (IBM, 2021). IBM Food Trust is a blockchain designed to make the food supply chain as transparent and trusted as possible. Developed by the American corporation IBM, this blockchain is based on the principle of an immutable ledger that contains information about the entire production process of food products. As a result, from the very beginning of production and ending with the table of the consumer, all the parties involved in the process, namely, farmers, processors, distributors, and retailers, as well as consumers themselves, can enter the necessary data regarding their food, such as the place of origin, details on processing, transport conditions, and even certificates. This information about production processes enables companies to demonstrate their commitment to sustainability and their responsibility toward this issue, something that is of great importance today from an environmental and social perspective. IBM Food Trust plays a critical role in improving transparency, traceability, and sustainability in the food supply chain, ultimately contributing to safer, more ethical, and sustainable food systems.

IBM has partnered with key companies like Walmart, Maersk, Bumble Bee Foods, Nestlé, Carrefour, and more to improve food safety, reduce waste, and enhance supply chain transparency. Walmart has been actively testing the use of blockchain in its food supply chain. Together with IBM, Walmart did several pilots to trace food products from the farm to the shelves at McDonalds. This move helped improve traceability and reduced the time taken to trace the contaminated food back to source (Kamath, 2018). One of the world's largest shipping companies in containers Maersk collaborated with IBM to create a blockchain platform known as TradeLens for global trade. This platform has helped to digitize supply chain processes, reduces the documentation effort, and creates real-time visibility in the movement of goods which creates value for shippers, freight forwarders, and customs authorities (Benton et al., 2018). Bumble Bee Foods also worked with SAP and IBM to set up a blockchain to trace yellowfin tuna from ocean to store. The process assists in ensuring transparency and accountability in the supply chain and fights illegal fishing and human rights abuse (Howson, 2020). Nestlé is one of the biggest players in food and beverage; it has been testing piloting blockchain to try and improve the transparency of the food chain. It has so far tested on coffee, palm oil, and milk, with an aim to improving sustainability and consumer food safety question (Mukhamedova and Mukhamedova, 2023). French retailer Carrefour allows customers to trace the origin of their meals by scanning their food label using a QR code. The code displays the entire history of the products, including origin points, transportation methods, harvest times, organic certifications, and more (Chang et al., 2021).

The manufacturing giant Unilever has already been part of and participating in various blockchain projects and collaborations attempting transparency and increased productivity throughout its supply chain. They have tried various pilot projects and initiatives funded by IBM, SAP, and other technology providers involving blockchain in supply chain. De Beers is a diamond giant that uses blockchain technology (Tracr) to trace diamonds from the mine to the consumer. Tracr ensures ethical sourcing of diamonds, and it assures the consumers they own genuine diamonds (Kshetri, 2022). Ford has also been testing blockchain supply chain to trace the supply of cobalt used in EV mostly sourced from the Democratic Republic of Congo ensuring the cobalt is ethically sourced (Kshetri, 2022).

The Intel Transparent Supply Chain blockchain is an opportunity option as opposed to the centralized model of trust among trusted suppliers and trusted manufacturers. A distributed application (DAPP) based on Ethereum smart contracts is utilized to allow supply chain participants to store platform-level data in the immutable blockchain (Intel, 2019). LVMH Moët Hennessy Louis Vuitton, a multinational conglomerate, uses blockchain to solve some of the challenges in the luxury industry like authenticity, responsible sourcing, and of course sustainability. The company employs a blockchain platform called AURA, which conducts authenticity checks for every luxury product and offers consumers detailed information about each of them as they purchase from its source of manufacture to its appearance on the counter (LVMH, 2021; Aura Blockchain Consortium, 2022). United Parcel Service uses blockchain as well in its supply chain processes. The company has partnered with technology companies like Blockchain in Transport Alliance (BiTA) for blockchain tests in the supply chain process (Supply Chain Magazine, 2020).

Provenance (PROV) is one of the blockchain projects that actively contribute to increasing the level of transparency and traceability in several industries. The primary focus of this project is the mission to facilitate access to secure records of the product's origins, certifications, and sustainability practices for consumers (Provenance, 2021). The platform allows companies to record, securely store, and share crucial information about their products. It includes the origin of the ingredients, the process of farming and processing, and the way food is delivered. Provenance makes it possible for consumers to access this information and makes them more likely to trust what they eat. It also has partnership with some of the companies in fashion industry, where companies demonstrate their sustainability and involve ethical ways to produce clothes by developing a product with a common record that cannot be corrupted. The electronic sector has a problem with conflict minerals, counterfeits, and disposal (Prendergast and Lezhnev, 2009). Therefore, Provenance helps the industry to track the source of the electronics, the presence of conflict minerals, and the level of environmental contamination. Chronicled, as a pioneering blockchain company, focuses on applying blockchain technology to the life sciences industry using the MediLedger network. They try to offer innovative solutions to the supply chain challenges by combining their blockchain expertise with deep industry knowledge, with a long-term vision to enable the life sciences community to develop MediLedger infrastructure, enhancing industry-wide innovation and ultimately solving related problems (Chronicled, 2021).

The VeChain blockchain platform covers multiple features and solutions that can address the complexities and issues arising in different businesses across diverse fields (VeChain Foundation, 2021). One of the great concepts to understand about VeChain is data interoperability, which makes it possible to securely interact and exchange data with other blockchain systems and technologies, and consequently enables the platform to integrate across different industries and ecosystems. The Vechain network offers numerous applications in many fields, such as a solution to deal with counterfeiting in luxury goods, traceability solutions in food industry considering the safety of food products from farm to table, monitoring CO_2 emissions and many more. To provide maximum effectiveness, the VeChain platform uses such cutting-edge technologies as the IoT, artificial intelligence, and big data analytics.

IoT enables real-time data collection and share them from physical objects to the blockchain. This integration makes a secure tracking of products through the supply chain and improves transparency and trust between parties. By using artificial intelligence in the accumulation of data and analysis, the efficiency multiplies, overalls and errors in logistics are minimized, and work capacity is increased due to optimal offers. From a sustainability perspective, VeChain's platform makes it possible to implement all eco-friendly practices and initiatives in the supply chain. It allows for monitoring the companies' sustainable sourcing and verification, assessment and reduction of renewable energy consumption and utilization, material waste, and processing waste. Within blockchain technology, companies can document and report the aforementioned measures, empower the corporate social responsibility pursued by organizations, and support the needs of modern-day consumers and investors who demand the corporate supply chain to be sustainable. Given the advanced analytics platform and prepared solutions for each sphere of the industry, VeChain offers all organizations the opportunity to pursue transparency, efficiency, and sustainability in all realms of practice.

Ambrosus is another specialized blockchain-based platform that addresses most of the challenges faced by supply chain management, focusing on the food industry and pharmaceuticals (Ambrosus, 2021). A distinctive feature of the platform is that at each stage of the supply chain, it composes an ample amount of data on the origin, quality, and conditions of product movement. Due to blockchain technology, the data are completely transparent, unchangeable, and secure. All information from the production to the sale of the final product is safely stored in the system. For example, in the food industry, the platform traces down any crucial item, starting with the ingredients of the product, all the ingredients and storage conditions, and the delivery route. The development of this amount of detailed information allows the manufacturer and the end consumer to verify the product's authenticity and quality. In the same way, Ambrosus works in the pharmaceutical industry. It accompanies the entire lifecycle of the shipment of pharmaceutical goods and ensures the compliance of all products with quality requirements and regulatory authorities. As a result, due to the secure nature of blockchain data, geographical coordinates, proof of adequate conditions, and any other data determining the quality of production are preserved. Overall, Ambrosus aims to develop transparent supply chains through which immutable data is transferred. The system also allows for proper sustainability reporting by each company and demonstrates to what extent the company complies with ESG principles and its range of interaction with ecology.

8.4.3 Blockchain Adoption by Sustainability Leaders

Several organizations and authorities that are active in maintaining and promoting sustainability are gradually expanding their capabilities in using blockchain in order to achieve their purpose. A few to mention are the United Nations, EU Digital Product Passport (DPP), Energy Web Foundation, World Wide Fund for Nature, and more.

The United Nations Environment Program (UNEP) explains in their whitepaper how blockchain advances the environmental sustainability as a tool and illustrates how it can solve environmental issues. Initiative activities, such as using blockchain for transmitting clean energy, or low-carbon development and mitigation of climate change, are pointing to the capacity of blockchain to facilitate effective actions and a better future with blockchain (UNEP, 2022).

The EU Digital Product Passport (DPP) is a digital document that keeps all the information about a product throughout its existence (BCG and Arianee, 2023). The main purpose of a passport will be to ensure transparency, sustainability, and circularity of a product in the EU. This document includes all the necessary information about a product. It makes it possible for anyone to find a country of origin, materials used for production, energy and waste, durability, and end of life information. DPP uses blockchain to store all this information, with the open access through scanning a QR code.

The Energy Web Foundation (EWF) was established in 2017, by Rocky Mountain Institute (RMI) and Grid Singularity, to accelerate the global transition toward clean energy. They created the Energy Web Chain, the first open-source blockchain platform specifically designed to support reliable tracking and sustainability application. The following are examples of this sustainability application: renewable energy certificate marketplaces, electric vehicle decentralized charging, and grid suppleness (Energy Web Foundation, 2021).

World Wide Fund for Nature (WWF) is an international non-governmental organization that was inspired and founded in Switzerland in 1961. WWF is the most famous environmental preserve organization worldwide. These days, it is active in around 100 countries and has almost five million members. The activities of WWF are very diverse, which is logical, as there are many problems on a global and local level. In addition to all the above, this company is also a pioneer in the use of blockchain and innovative technologies, such as a blockchain-based solution for tracking seafood. This solution makes it easier and more transparent to trace the provenance of seafood supply, which helps solve problems such as fishery poaching and unsustainable harvesting. Also, tracking and recording wildlife movements can be done with blockchain. Recording this information will allow environmental protection organizations to have clear tracking and minimize the possibility of information loss or fraud. And lastly, in carbon footprint reduction, blockchain enables greater transparency and reduced fraud and less restricted access to global markets. This is possible through tokenization. Tokens are a method of changing real assets into digital units that are on the blockchain. Each token stands for a certain quantity of carbon absorbed or prevented from being emitted. These tokens digitally authorize companies and individuals worldwide to trade the quantity of carbon they have absorbed or diminished (World Wide Fund for Nature, 2023).

8.5   THE FUTURE OF BLOCKCHAIN IN SUSTAINABILITY

In conclusion, the integration of blockchain technology into the sustainability reporting sector is a significant achievement toward implementing an authentic and reliable environmental accountability process. On the one hand, organizations have employed blockchain to correct the inadequacies of traditional reporting methods, a situation that has seen them benefit from, among others, increased traceability and guaranteed lifetime records. On the other hand, it is true that blockchain has not been given the priority it deserves. To achieve this, several strategic initiatives could be considered the following.

While a significant emphasis and purpose unite the blockchain projects, the first problem that can be pointed out is the complex nature of blockchain, which requires up-to-date technical knowledge as well as major changes in infrastructure, which may be difficult for companies (Crosby et al., 2016). To overcome this issue, education could be one solution. Setting up various workshops and seminars hosted by developers and activists in this field can inform business leaders of its importance, benefits and practical applications in sustainability reporting (Smith and Lee, 2020). One of the most useful solutions can be started from schools and universities. In order to institutionalize this issue in the society, we must start training the next generation of business leaders from today, and this is possible through educational institution (Johnson and Turner, 2023). Another solution can be a partnership with leading technology companies in the use of blockchain. Fortunately, blockchain has the ability to be developed or even custom designed. By directly partnering with the implementing firms, it is possible to consider the company's expectations and ambiguities and finally reach a tailored blockchain platform (Lee and Nguyen, 2022). In general, organizing seminars, workshops and conferences, and partnering with the key organization in the area of sustainability reporting, blockchain community will help potential users dispel several their misconceptions and acknowledge a number of their abilities to integrate blockchain for sustainability reporting.

Another problem that the use of blockchain may face is the high energy consumption in some blockchain systems, which contradicts the main nature of sustainability (Krause and Tolaymat, 2018). Traditional blockchain systems like proof of work (PoW) are stigmatic energy-intensive, since requires a huge computational power to solve complexity mathematical equations. Therefore, the more power-usage, the more electricity it consumes. Consequently, this aspect contradicts the general nature of sustainability that blockchain is to support. Fortunately, apart from this method, there are other solutions for this issue that will be mentioned briefly in simple terms.

Proof of stake (PoS): This method is similar to owning shares in a company. Suppose you are a shareholder of a company and you are involved in the company's decisions as much as your share. Based on the amount of digital currency, which is locked as a stock, and the period of holding, the more chance of being selected to approve transactions (King and Nadal, 2012).

Delegated Proof of Stake (DPoS): The concept of this method is similar to voting in an election. All the digital currency holders can vote for presenters or delegates, who validates the transactions on behalf of them (Larimer, 2014).

Proof of authority (PoA): Proof of authority is like a little administration that supervises everything. The validation of the system is ensured by a fair number of experts who have a good reputation and trustable. Such individuals must guarantee specific rules that the network is both safe and transparent (De Angelis et al., 2018).

Although this problem can be improved with alternative methods as mentioned, each of these methods has its own advantages and disadvantages, and their selection requires specific conditions (Miglani et al., 2020).

It is also worthy to mention that the blockchain has a short background, means the subject is fresh and evolving day by day, so the regulations to this matter are not in a solid structured framework yet (Iansiti & Lakhani, 2017). Another issue that can be challenging is excessive transparency in published information. Since blockchain enables free access to the recorded information, which can be sensitive and raises privacy concerns (Pilkington, 2016). These issues could be addressed by regulatory compliance, ensuring that blockchain platforms follow specific established frameworks, to avoid possible legal problems that could arise from concerns about the disclosure of organizational data privacy or other related challenges.

In connection with the increasing concerns toward data anonymization, blockchain could be regulated under the specific rules like General Data Protection Regulation (GDPR)[2] in Europa or other similar laws across the world. However, GDPR provisions as such the right for modifying or deleting data under some circumstances, is directly in conflict with the nature of the blockchain (Read and Pehlivan, 2020). Blockchain developers must find a solution to this aspect that can both solve this issue and not violate the principle of incorruptibility and immutability of blocks (Stach et al., 2022).

8.5.1 SUSTAINABILITY REPORTING AND TRANSPARENCY: INDIGENOUS EXPERIENCE CASE STUDY AND THE POTENTIAL FOR BLOCKCHAIN TECHNOLOGY

The potential for blockchain technology application in sustainability reporting can be illustrated by reference to the Indigenous experience(s) in the extractive industries. Similarly, other stakeholders – corporate proponents from all sectors of resource development, consultants of all stripes associated with the extraction industry, governments at all levels, and individuals, communities, and businesses located along the supply chain of distinct extractive enterprise(s) could arguably benefit from its application and use in assessment, monitoring, and reporting during the life of the extractive industry project. This is because the inherent attributes of blockchain technology, including immutability, decentralization, and especially transparency are effectively those sought after by stakeholders seeking assurance, accuracy, and confidence in sustainability reporting. The Indigenous stakeholder in particular has historically experienced profound negative impacts from extractive resource activity. Most insidious is the impact that extractive industry operations have on the environment that is home to Indigenous people. In many cases the social and cultural fabric of Indigenous community that is tied to traditional territory and resources has been deleteriously impacted. Today, attempts to mitigate negative impacts are negotiated and included in benefits agreements between proponents and Indigenous stakeholders. Additionally, companies within the industry who are concerned with the social license to operate, and with self-regulation are slowly including Indigenous participation in certification programs focused on industry self-regulation. From the Indigenous perspective the results of these efforts have been mixed pointing to a key shortfall in Indigenous involvement in the oversight, monitoring, and sustainability reporting related to negotiated agreements and certification programs. Complementary to Indigenous involvement in project oversight and reporting is the application of appropriate blockchain technology that would engender confidence in the accuracy, and transparency in the reporting that Indigenous people, and by extension, society seeks. Adoption of these two critical recommendations would enhance trust and confidentiality where appropriate, and transparency as required, in sustainability reporting. Highlighting these elements through a summary review of a contemporary case study from the extractive industries could bring these elements into focus.

8.5.2 Indigenous Involvement With the Extractive Industry: Trans Mountain Pipeline Expansion Project (TMX) Case Study

One such case study is the Trans Mountain Expansion Project (TMX) in Western Canada. TMX is an extractive industry project where there has been a small measure of Indigenous Local Knowledge (ILK) input through negotiations leading to a number of signed Mutual Benefit Agreements (MBA) between First Nations and TMX, and where the potential exists for ESG reporting that reflects to some extent Indigenous concepts of sustainability. Arguably, TMX legally mandated sustainability reporting requirements situate it as an ideal candidate for blockchain technology utilization in its reporting process that can mitigate the many key issues that dog the contentious megaproject. From its outset, the project was plagued with disagreement and protests, court challenges, and political intervention (Doebeli and Giang, n.d.). What follows provides a brief contextual description of TMX with a focus on the Indigenous stakeholders involved. The description will highlight at a high level some of the contentious aspects of the TMX–Indigenous relationship where blockchain technology's inherent characteristics of immutability, decentralization, and transparency could, if applied with the involvement of the Indigenous stakeholders generally, and ILK knowledge holders specifically, transform TMX's sustainability reporting into one that promotes accuracy, provides assurance, rebuilds trust, and inspires confidence.

The Trans Mountain Expansion Project is a major oil pipeline expansion project that traverses the provinces of Alberta and British Columbia, Canada's two westernmost provinces. Originating in Edmonton, the capital city of Alberta, and culminating at port in Burnaby, British Columbia, the project twins with the original Trans Mountain Oil Pipe Line Company pipeline to increase the transportation capacity of crude oil and semi-refined and refined oil. The 1,150-kilometer-long pipeline system, once completed and operational, will increase transportation capacity from 300,000 barrels per day to 890,000 barrels per day (Doebali and Giang, n.d.). The original pipeline that was approved in 1951 was unimpeded by public and environmental assessment processes, and amazingly, only took two years to construct from the time of approval (Doebali and Giang, n.d.).

That is in stark contrast with today's TMX project proposal and construction timeline. The twinning project was originally proposed in 2012 by Kinder Morgan, who owned the pipeline at the time. The project proposed to follow 73% of the existing pipeline route, utilize 13% of existing infrastructure corridors, and thus only needed to acquire 11% new right of way acquisitions. From its conception, the TMX pipeline twinning proposal encountered many challenges due to public consultation and environmental assessment requirements that had been legislated over the years since the original pipeline was constructed (Doebali and Giang, n.d.). Many of the new legal requirements reflected an increasing awareness of the environmental impacts of economic development. Hence, the project only received National Energy Board (NEB) recommendation for approval in 2019 after lengthy and protracted consultations, protests, and court challenges. Following the NEB recommendation, the government of Canada ordered certification of the project, also in 2019. The project officially launched on 1 May 2024, approximately four and half years after it was approved, and approximately 12 years after it was originally proposed (Doebali and Giang, n.d.). What had changed in the approval processes in the intervening years between the construction of the original pipeline and the planning and construction of the Trans Mountain Pipeline Expansion Project was growing societal concerns over the impacts of extractive industry activity on the environment and society generally and the overall economic sustainability of such large-scale development. Accompanying these concerns was the growing body of legal precedent in support of expanded Canadian Constitutional recognition of Aboriginal rights and title in Indigenous traditional territories. Pursuant to now-established law governments in Canada have a legal duty to consult with Indigenous peoples and cannot unjustly abrogate their rights and title interests (Doebali and Giang, n.d.; Gibson and O'Faircheallaigh, 2015). Canadian provincial and federal governments formally fulfill this duty through their respective environmental assessment agencies. In the case of TMX, the NEB was charged with fulfilling the role of "consultation and accommodation" negotiations through its legally mandated processes and procedures. Further strengthening Indigenous claims to traditional and ancestral territory occurred when Canada became a signatory to the United Nations Declaration on the Rights of Indigenous Peoples (UNDRIP) in 2016, which requires that Canadian governments obtain "free, prior, and informed consent" from Indigenous nations that are potentially impacted by projects that propose to infringe upon their resources and territories (Doebali and Giang, n.d.; United Nations, 2008). Therefore, government policy reflecting public concern with environmental impacts, the beneficial nature of obtaining a social license to operate, and national and international law developments with respect to Aboriginal and Indigenous rights and title motivated, and compelled, TMX to enhance its consultation processes, which in turn critically expanded the approval process timeline.

The Trans Mountain Expansion Project proposed to traverse the traditional territories of 133 Aboriginal peoples that included some Metis nations and many First Nations. Forty-three (43) of these Nations signed MBAs (Doebali and Giang, n.d.). The comparatively small number of signed MBAs, some of which were signed under protest, reflects a variety of views regarding support and opposition to the project. From the Indigenous perspective, several issues prevail that influence their perspectives regarding the TMX project. Among these, three key issues pertinent to sustainability and sustainability reporting stand out including: the experience of Indigenous people in the Indigenous–settlor relationship, the previously unfettered activities of extractive resource industry that results in socio-cultural, and environmental degradation, and the power divide that exists between industry players and weaker societal players, including Indigenous groups, that results in extractive industry "others" commandeering definitions, operations, and monitoring of sustainability and sustainability reporting (Ostrowski, 2020). Governments in Canada continue to support projects like TMX because it is in the national interest to do so despite much stakeholder opposition and resistance, pointing to the ongoing colonization object of Indigenous–settlor relations, which in its essence pits the individualist nature of capitalism against the collective nature of Indigenous societies, and "man's domination over all others" (Ostrowski, 2020). The secretive nature of the extractive industry pre-21st century that invoked perceptions of corruption and collusion in the colonial and neo-colonial exercise continues to fuel feelings of distrust to this day (Ostrowski, 2020). And the minimum participation of Indigenous representatives in the ongoing development of environmental assessment processes, or extractive industry standard creation, and the lack of enforcement of contractual MBA monitoring requirements points to a lack of consideration and respect for the knowledge and ways of knowing about sustainability that affected Indigenous nations and Indigenous knowledge keepers have to offer. The historical significance is that a consistent "air" of distrust persists among impacted Indigenous stakeholders that is furthered by a lack of transparency in sustainability reporting in the extractive industry that at best only meet minimal unilaterally established governmental, industry, and arguably weak societal standards (Ostrowski, 2020). Hence, it is not surprising that from the beginning TMX faced protest and litigation from Indigenous stakeholders, which continues to this day.

Broadly speaking, there are dual complementary solutions that could reverse this trend and work to create trust between and among governments, Indigenous stakeholders and extractive industry proponents. One solution rests on the notion that Indigenous participation in the creation of laws related to environmental assessment processes, and in the formation and establishment of extractive industry certification standards, would provide space for Indigenous technical knowledge, ways of knowing, and understanding of sustainability to inform the project from start to finish. The second related solution is a recommendation for the application of blockchain technology for the record keeping, monitoring, and sustainability reporting of the project that could provide accuracy and assurance – hallmarks of transparency that would build trust and confidence among the parties in the relationship. The sheer number of Indigenous Nation stakeholders with varying perspectives about extractive industry projects that TMX had to consult and accommodate with, and now have to report to, demand the transparency and efficiency in sustainability reporting that blockchain technology has to offer. From TMX's perspective this ought to be an "a priori" assumption given the national and international recognition of Aboriginal and Indigenous rights in play. This would positively impact the projects legal reporting obligations, its social licence to operate, and its (and by proxy, Canada's) reconciliation efforts with Indigenous stakeholders. These two proposed solutions are elucidated later.

Canada has been populated by Indigenous people for millennia. In every corner of Canada, Indigenous people have come to know the land and all that live in body and in spirit on the land that we refer to as "Mother Earth" and as "Turtle Island". Throughout time, the landscape of Indigenous territories has changed in many ways due to the forces of nature. In the face of these changes, Indigenous ways of knowing and localized knowledge of, and experience with the environment has been key to Indigenous adaptation and survival. The combination of "ways of knowing" and local knowledge is fundamental to and informs Indigenous peoples' concepts of "sustainability". And, while this is obvious to Indigenous Peoples, it has only recently been substantively recognized in other forums. One such entity is the Intergovernmental Platform on Biodiversity and Ecosystem Services (IPBES), which

> recognizes that the diverse social, cultural, and environmental knowledge of indigenous peoples and local communities (IPCL) contributes extensively to sustainability across large parts of the globe, and thus has a major role to play in assessments and policy formulation for biodiversity and ecosystem services.
>
> (Hil et al., 2020)

IPBES' focus specifically on Indigenous and Local Knowledge (ILK) concludes that "Recognizing, respecting, and engaging with humanity's diverse knowledge systems can help secure the future of nature and natures' linkages with people" (Hill et al., 2020). This speaks well for nature and people, and arguably, could improve corporate sustainability as well.

Unfortunately, Indigenous experience with the extractive industries and companies like TMX does not reflect this conclusion. There is scant evidence that supports recognition of Indigenous ways of knowing and the existence of Indigenous technical knowledge outside of that which comports with contemporary scientific knowledge, and where regulatory regimes compel extractive industry operatives like TMX to take into consideration and account for. And yet, the extractive industries are changing Indigenous home landscapes in ways not known prior to colonization. The history of extractive industry players and Indigenous Peoples' relations is characterized by self-interest on the part of industry, accompanied by considerable negative impacts on the environment, and on the social and cultural fabric of the communities and the territories within which they operate (Sosa and Keenan, 2001). ILK is brushed aside adding to the distrust created by years of unrestrained practices of the extractive industry firms, and the broken treaty and political promises of successive federal and provincial governments (Meadows et al., 2019). Still, opportunities to reverse this trend may now be foreseeable. For now, ILK still exists. And there may be time to acknowledge its potential contribution to positively affect extractive industry practices. IPBES describes ILK as "dynamic and holistic", "highly diverse", and "managed by distinctive cultural institutions" (Hill et al., 2020). At the least these characteristics acknowledge adaptive capacity for the relationship between people and nature through time and space, and for the involvement of knowledge keepers in sharing ILK (Hill et al., 2020).

Practically speaking, this recognition is a call for participation of the Indigenous knowledge holders in assessment, monitoring, and reporting that ties with Indigenous concepts of sustainability, and, by extension, where additional concepts of societal and corporate sustainability may also be in play. The potential for Indigenous participation and inclusion of ILK, including Indigenous concepts of sustainability, and sustainability reporting in the extractive industries arguably exists in current practices of negotiating benefits agreements (O'Faircheallaigh, 2020), and in recent trends in the development of industry self-regulation certification programs (Meadows et al., 2019). Meadows et al. provide further support for the proposition suggesting that "Best practice design, implementation and verification design of sustainability standards underpinning certification programs requires active participation of Indigenous Peoples" (Meadows et al., 2019, p. 1). Seizing this existing potential could rebuild trust and enhance transparency that could temper and mitigate the negative impacts of the extractive industry on Indigenous people.

Contemporaneously with developments in Canadian and International laws and agreements related to the environment and Indigenous peoples of Canada, as well as international accords and laws, and public pressure on governments and business to think beyond the profit motive and, to reflect societal concern for the people and the planet, companies like Trans Mountain Pipeline and others in the extractive industries are compelled to apply for and receive environmental assessment certificates, among other licenses and permits, before construction of the project can begin (Gibson and O'Faircheallaigh, 2015). The application process for major projects is often extensive and requires considerable technical detail reflecting intensive research in support of the project's viability. Project proposals that are interprovincial in nature, and that fall under jurisdictions of both Federal and Provincial governments, such as the TMX pipeline project, make application to both agencies in charge of the environment. Usually, companies will ask to harmonize dual government ministry processes so as create efficiencies. Both levels of governments often honor these requests, with one government or the other taking the lead (Gibson and O'Faircheallaigh, 2015). Given the interprovincial nature of the TMX project, the National Energy Board (NEB) took the lead (Doebali and Giang, n.d.). Applications are reviewed by the responsible ministries and once cleared, invitations are sent out to potentially affected or impacted stakeholders. The TMX project impacted many stakeholders, traversing as it proposed to do, across two provinces in a continuous line from Edmonton, Alberta to the Westridge Terminal in Burnaby, British Columbia. Environmental assessments require proponents to consult with all stakeholders it impacts. And, in many cases, proponents enter into formal arrangements with stakeholders that can involve restitution for the impacts created by the project, including accommodation in the form of financial compensation where appropriate. Most importantly, the arrangements reflect the proponent's motivation to obtain a societal license to operate from impacted stakeholders such as communities along the proposed route (Gibson and O'Faircheallaigh, 2015). Unique and key among the stakeholders impacted by the TMX proposal were the 133 Indigenous Nations mentioned earlier. Since the environmental assessment application proposed to construct the pipeline through the traditional territories of First Nations, and in many cases, through the Indian reserves set aside for the First Nations people, Canada had a duty to consult and accommodate the Indigenous Nations pursuant to established Canadian law, and more so, to seek the "free, prior and informed consent" of the Indigenous stakeholders in accord with international law before approving the TMX project (Gibson and O'Faircheallaigh, 2015). As often happens in these cases Canadian governments, in exercising their jurisdiction, leave the "accommodation" aspect of the duty to consult to project proponents. Hence, it was left up to TMX to commence negotiations with affected Indigenous nations to discuss accommodation details (Doebali and Giang, n.d.). Negotiations, if successful, result in agreements between the proponent and the stakeholder called Mutual Benefit Agreements (MBA), or Impact Benefit Agreements (IBA), or sometimes Reconciliation Agreements (RA).

Indigenous critics of this process highlight various high-level issues with the process. First, there is no requirement for project proponents to consult with Indigenous stakeholders prior to submitting environmental assessment (EA) applications, and thus, to consider, Indigenous knowledge about potential impacts on its territory, people, and way of life in advance of proposal submission. Second, and perhaps most importantly, Indigenous stakeholders have no input on the development of EA processes, and thus, these processes in no way reflect fundamental ILK or Indigenous notions of sustainability (Meadows et al., 2019). Proponents like TMX first apply for environmental assessment office certification to relevant federal and provincial ministries. Only then are affected First Nations notified of potential infringements on their rights and title interests. And only once notified through this process are First Nations able to respond to a proposal. The Indigenous response usually reflects their Indigeneity and their understanding of sustainability of their traditional land and resources. The Indigenous response is then reviewed by the government agency and eventually, government officials who do not understand the Indigenous perspective determine the validity of the Indigenous response. Third, Indigenous stakeholders are "ab initio" forced into a weak negotiation situation where they are compelled to respond to sophisticated applications. Since most First Nations lack financial resources and other capacity to respond effectively to technically rigorous proposals like the TMX pipeline proposal, they must seek this assistance from governments who often support the projects, and from the proponents themselves who, as part of the process and cost of doing business, will provide capacity assistance to the Indigenous stakeholders (O'Faircheallaigh, 2020). Indeed, this was the case with TMX, when Canada became owner of the mega project (Doebali and Giang, n.d.). Arguably, Canada, as the new owner of TMX, created a conflict-of-interest situation. The reality of these issues highlights the injustice of the situation and signifies the need for ongoing transformation with governmental processes, extractive industry standards for consultation as part of the process of gaining a social license to operate, and the inclusion of ILK and Indigenous participation in all areas requiring change.

As with all things worthwhile achieving these changes will take much time, will and effort, and resources. In the interim, there are small openings in the current system for promotion of IDK. At the governance level, First Nations are exercising their self-government authority to create First Nations environmental assessment processes built upon Indigenous values and Indigenous knowledge of the interconnectedness of the land and the resources. Squamish Nation is one, for example, that has created such a process. Other First Nations will do the same as more self-governing authority is realized through modern treaty-making processes. Indigenous groups across Canada have been initiating resurgence efforts toward re-establishing teachings and values related to traditional land, people, and resources. Canadian governments, in their own efforts at reconciliation, are acknowledging and supporting Indigenous initiatives in economic and resource development. And, as resurgence and reconciliation efforts increase so will the negotiation leverage of Indigenous groups increase at the "consultation and accommodation" tables where Indigenous stakeholders and industry proponents meet to discuss industry standards and to negotiate MBAs.

With respect to negotiated agreements, O'Faircheallaigh makes the case that "IBAs have considerable potential to effectively monitor project impacts and delivery of project benefits, but a broader analysis of IBAs shows that this potential is often not realized" (O'Faircheallaigh, 2020, p. 1338). He attributes non-realization of this potential on the lack of support by government, a lack of financial resources to implement monitoring and reporting processes, and a dearth in Indigenous organizational resources (O'Faircheallaigh, 2020). In reaching this conclusion, however, he posits that Indigenous parties can be active in influencing and implementing monitoring arrangements written into IBA agreements (O'Faircheallaigh, 2020). Most importantly, O'Faircheallaigh states "The active involvement of Indigenous peoples is in fact critical because they hold cultural and environmental knowledge that must be mobilised if monitoring is to be effective" (O'Faircheallaigh, 2020, p. 1338). The increased sophistication of First Nations government noted earlier and the increased leverage held by Indigenous groups when extractive industry activities are located in Indigenous traditional territory suggests that they are in a position to negotiate terms in agreements that could reflect ILK. Furthermore, IBA terms could provide for accommodation in the form of sufficient funding from extractive industry proponents to support monitoring systems and to establish reporting processes, including sustainability reporting that is inclusive of meaningful Indigenous participation. In doing so, extractive industry companies like TMX could effectively demonstrate fulfillment of societal and Indigenous demands for accurate and transparent sustainability reporting not just to government agencies but to directly affected parties.

TMX negotiated 43 MBAs with Indigenous Nations (Doebali and Giang, n.d.). TMX–Indigenous MBAs are confidential to the signatories and so the specific details of each agreement are not publicly available. Generally, however, MBAs include provisions that are far ranging and "which seek to shape the occurrence and distribution of costs and benefits arising from major projects,. . . and which embody the support of Indigenous entities (landowners, communities, governments) for the project concerned" (O'Faircheallaigh, 2020, p. 1339). Some may include provisions for monitoring and sustainability reporting that goes beyond that required by EA legislation (Gibson and O'Faircheallaigh, 2015). O'Faircheallaigh identifies two dimensions to monitoring provisions in MBAs. He identifies an internal dimension whereby data gathered, and periodic strategic operational review is conducted and reported on to determine compliance with the agreement's intent (O'Faircheallaigh, 2020). "The 'external' dimension relates to the role of the agreement in monitoring the impact of the project on people, society, culture, and the environment" (O'Faircheallaigh, 2020, p. 1339). Assuming that Indigenous parties to the TMX agreements have maximized their leverage and negotiated project agreements informed by ILK through the life of the project, the logical next step is to ensure that monitoring and reporting are conducted using tools that maintain the confidence of the agreements, provide transparency, and build trust in the partnership.

Having made the case for the relevance of Indigenous inclusion in extractive industry standards development in the absence of Canadian government's desire to be involved beyond minimal EA-regulated efforts, and Indigenous stakeholder participation in projects like TMX from planning to decommissioning it as important to address the critical question of how extractives can transparently report on the sustainability of a project while continuing to involve Indigenous participation in the reporting process. Arguably, the application of blockchain technology can play a significant role in recording, monitoring, and reporting in a way that simultaneously maintains confidentiality where that is required as with MBAs and realizes transparency that is demanded by all stakeholders, particularly Indigenous stakeholders. The mechanism for reporting is important given the historical Indigenous mistrust of government and extractive industry players who customarily hold the pen on sustainability reporting. Blockchain technology creates the opportunity for Indigenous peoples to also hold the pen so that they can critically analyze and add commentary reflective of Indigenous technical knowledge to sustainability reporting. Immutability, decentralization, and transparency are inherent characteristics of blockchain technology that supports this proposition. Reporting based on these attributes would instill confidence in the accuracy of the reports because inputs can be more readily fact checked and verified by third parties as necessary.

Earlier, the background and context for sustainability reporting, and building the case for a role for blockchain technology are provided in sustainability reporting. Building on prior research, it can be concluded that stakeholder participation could be increased, third-party verification improved, and that technological capacity building and training rendered more effective with the use of blockchain technology and conclude that "Generally, the blockchain is an appropriate tool for the recommended strategies in terms of making sustainability reporting more reliable and transparent". Furthermore, this chapter in reaching the conclusion finds that there exist numerous frameworks and standards for sustainability reporting that industries and companies can employ to meet stakeholder expectations. Prominent among these standards are Global Reporting Initiative (GRI) and United Nations SDGs, which when aligned with the authors' proposal to employ blockchain technology to transparent sustainability reporting, would go a long way to establishing trust and confidence among stakeholders, and to re-establishing trust and confidence where that has been lost. Arguably, the TMX project that has become the Trans Mountain Corporation (TMC) is one company in the extractive industries that could realize this were it to pursue more vigorously its goal to be more transparent. The TMC ESG report of 2023 states its desire to "be transparent about our practices and performance. The goal of this ESG report is to communicate the environmental, social, and governance initiatives and key metrics that demonstrate our progress to date and our commitment to continual improvement" (Trans Mountain Corporation, 2023, p. 12). TMC ESG reporting is based upon stakeholder-influenced ESG priority topics and select standards drawn from the Sustainability Accounting Standards Board (SASB), the Task Force on Climate-related Financial Disclosures (TCFD), the 17 UN Sustainable Development Goals (SDGs), and other midstream company reports (Trans Mountain Corporation, 2023, p. 12). TMX senior management decided which ESG topics would be reported upon based upon company impact and stakeholder interests. Presumably stakeholder interests were gathered and assessed during the regulatory mandated consultation processes, and in the case of Indigenous stakeholders, according to terms of negotiated IBAs. Considering that there continues to be opposition to the project not all stakeholder interests are likely reflected in senior management's ESG chosen report topics. Furthermore, senior management could change the ESG standards in the future according to their perception of stakeholder interests at the time, and of course, according to how the standards and the ESG reporting impact the success of the company. And considering that TMC holds the pen on its reporting then stakeholders, including Indigenous stakeholders, could reasonably question TMC sincerity about its stated goals of providing transparent sustainability reporting, and continual improvement. Clearly, from the Indigenous perspective, there is need for input on the substantive ESG standards selected for reporting so that standards are inclusive and cannot be unilaterally changed, and for the reporting process itself to be transparent.

Elsewhere in its ESG 2023 report TMC states that it has promised to establish an Indigenous committee and that it has not accomplished that to date (Trans Mountain Corporation, 2023, p. 7). It does state it continues to work on that. With that promise in mind TMC could work with Indigenous stakeholders to incorporate ILK and Indigenous notions of sustainability to its standards regime as noted earlier, and to add its progress in that area to its sustainability reporting. Together, Indigenous stakeholders and TMC, and other stakeholder could adapt blockchain technology to discuss and arrive at sustainability standards that for the record are immutable and decentralized, ensuring mutually beneficial interaction before any changes can be enacted to the substance and measure of sustainability. Applying blockchain technology to other aspects of TMC–Indigenous stakeholder relationship such as negotiated MBAs could ensure compliance in monitoring and reporting that can be verified by third-party professionals and Indigenous knowledge keepers, while continuing to maintain the confidentiality of the MBAs and established relationships. And finally, TMC could use the adapted blockchain technology in its overall sustainability and ESG reporting to the benefit of all key stakeholders. The resultant transparency could allay fears and mistrust of some if not all problematic and antagonistic stakeholders that continue to challenge TMC operations. This will improve its social contract position and more than satisfy regulatory compliance.

8.6   SUMMARY

This section of the chapter set out to illustrate the potential use of blockchain technology in sustainability reporting by highlighting the case of Indigenous stakeholders involved with the extractive industries. The Trans Mountain Expansion Project is instructive due to the many Indigenous stakeholders affected by the project. The negative experience of Indigenous peoples at the hands of colonial and Canadian governments who came to control Indigenous lands and resources is well documented. That history, coupled with the historically secretive nature of the extractive industry and the devastating environmental and social impacts of the industries' operations on traditional territories and peoples, has created considerable mistrust in the relationship among the three stakeholders. This mistrust is further exacerbated by minimal legally mandated environmental standards and potential non-compliance with impact benefit agreement provisions related to monitoring and reporting. Beyond issues of non-compliance with the provisions, extractive company operatives most often hold the pen on monitoring and reporting of their sustainability activities. This minimizes Indigenous input and participation to the negotiating stage of impact benefits agreements, and in some cases to monitoring of company operations. Extractive industry operatives are therefore free to report on their sustainability efforts in a manner that best suits them. When companies' reporting contradicts stakeholder perspectives about company operations and ESG related impacts, then questions about accuracy and transparency of sustainability reporting are raised. The extractive industry case of the TMX pipeline raises these issues and illustrates these challenges and provides a basis for solutions contemplating the potential for blockchain technology in incorporating ILK in establishing its industry and corporate ESG standards, in consultation and accommodation discussions and compliance with negotiated MBA/IBA/RAs, and in monitoring and reporting of mutually selected ESG standards. Thus, TMX sustainability reporting would be transparent, and work to re-establish trust that was lost, and gain new stakeholder trust in TMX efforts at gaining an expanded social license to operate.

NOTES

1 Concept of a 51% attack is a security threat to blockchain network where a single entity or group controls more than half of the network's computational power, potentially allowing them to manipulate transactions and compromise the integrity of the ledger.
2 The General Data Protection Regulation (GDPR) is a European Union regulation that sets guidelines for the collection and processing of personal information from individuals within the EU.

REFERENCES

Adams, C. A. (2004). The ethical, social and environmental reporting-performance portrayal gap. *Accounting, Auditing & Accountability Journal*, 17(5), 731–757.

Adams, C. A., & Frost, G. R. (2008). Integrating sustainability reporting into management practices. *Accounting Forum*, 32(4), 288–302.

Ambrosus. (2021). *Ambrosus: Blockchain-Powered Supply Chain Solutions, Report*. Ambrosus.

Aura Blockchain Consortium. (2022). *Aura Blockchain Consortium Launches Aura SaaS for Luxury Brands, Report*. Aura Blockchain Consortium.

Benton, M. C., Radziwill, N. M., Purritano, A. W., & Gerhart, C. J. (2018). Blockchain for supply chain: Improving transparency and efficiency simultaneously. *Software Quality Professional*, 20(3).

Boston Consulting Group (BCG) & Arianee. (2023). *The Case for Native Digital Product Passport Tokenization (Joint Report)*. Boston Consulting Group.

Broeder, L. (2024). *Becoming Green Together: The Case of SAENZ*. Delft University of Technology.

Brown, T., & Green, F. (2021). Blockchain's role in transparent sustainability practices. *Journal of Environmental Management*, 285, 112–119.

Cardu, M., Farzay, O., Shakouri, A., Jamali, S., & Jamali, S. (2023). Feasibility assessment of acid gas injection in an Iranian offshore aquifer. *Applied Sciences*, 13(19), 10776.

Chang, J. A., Katehakis, M. N., Shi, J. J., & Yan, Z. (2021). Blockchain-empowered newsvendor optimization. *International Journal of Production Economics*, 238, 108144.

Chronicled. (2021). *How to Supercharge Your Revenue Management System through MediLedger*. Whitepaper.

Clark, G. L., Feiner, A., & Viehs, M. (2015). From the stockholder to the stakeholder: How sustainability can drive financial outperformance. *Journal of Sustainable Finance & Investment*, 5(4), 210–233.

Clark, G. L., & Yavuz, E. (2018). The shifting politics of ESG. *Journal of Sustainable Finance & Investment*, 8(1), 1–20.

Crosby, M., Pattanayak, P., Verma, S., & Kalyanaraman, V. (2016). Blockchain technology: Beyond bitcoin. *Applied Innovation Review*, (2), 6–10.

De Angelis, S., Aniello, L., Baldoni, R., Lombardi, F., Margheri, A., & Sassone, V. (2018). PBFT vs proof-of-authority: Applying the CAP theorem to permissioned blockchain. In *Proceedings of the Second Italian Conference on Cyber Security (ITASEC 2018)* (p. 2058). Milan, Italy. CEUR workshop proceedings (Vol. 2058). CEUR-WS.

de Freitas Netto, S. V., Sobral, M. F. F., Ribeiro, A. R. B., et al. (2020). Concepts and forms of greenwashing: A systematic review. *Environmental Sciences Europe*, 32, 19.

Dillard, J., & Vinnari, E. (2017). Theorizing transparency: (De)legitimization through social and environmental reporting. *Organization & Environment*, 30(1), 62–81.

Doebeli, G., & Giang, A. (n.d.). *Open Case Studies: Trans Mountain Pipeline Expansion Project*. Retrieved July 3, 2024 from: https://wiki.ubc.ca/Documentation:Open_Case_Studies/ IRES/Trans_Mountain_Pipeline_Expansion_Project.

DP. (2019). *CDP Climate Change Questionnaire 2019: Guidance for Companies Reporting on Climate Change*. Organizational Report.

Eccles, R. G., Ioannou, I., & Serafeim, G. (2012). The impact of corporate sustainability on organizational processes and performance. *Management Science*, 59(5), 1045–1061.

Eccles, R. G., & Serafeim, G. (2013). The impact of corporate sustainability on organizational processes and performance. *Management Science*, 59(5), 1045–1061.

Energy Web Foundation. (2021). *Energy Web Chain: Innovations in Sustainability*. Rocky Mountain Institute & Grid Singularity.

Fathi, H. (2021). Sustainability reporting: A comprehensive overview. *Sustainability Accounting, Management and Policy Journal*, 12(6), 1132–1159. https://doi.org/10.1108/ SAMPJ-06-2020-0259.

Freeman, R. E. (2010). *Strategic Management: A Stakeholder Approach*. Cambridge University Press.

Freeman, R. E., Harrison, J. S., Wicks, A. C., Parmar, B. L., & De Colle, S. (2010). *Stakeholder Theory: The State of the Art*. Cambridge University Press.

Gibson, G., & O'Faircheallaigh, C. (2015). *IBA Community Toolkit: Negotiation and Implementation of Impact and Benefit Agreements*. Walter and Duncan Gordon Foundation.

Global Reporting Initiative (GRI). (2016). *GRI Sustainability Reporting Standards*. Global Reporting Initiative (GRI).

Hahn, R., & Kühnen, M. (2013). Determinants of sustainability reporting: A review of results, trends, theory, and opportunities in an expanding field of research. *Journal of Cleaner Production*, 59, 5–21.

Hahn, R., Reimsbach, D., & Schiemann, F. (2015). Organizational antecedents of environmental management practices. *Sustainability*, 7(7), 8559–8572.

Hahn, T., & Figge, F. (2011). Beyond the business case for corporate sustainability. *Business Strategy and the Environment*, 20(7), 437–452.

Hertwich, E. G., & Wood, R. (2018). The growing importance of scope 3 greenhouse gas emissions from industry. *Environmental Research Letters*, 13(10), 104013.

Hill, R., Adem, C., Alangui, W., Molnar, Z., Aumeeruddy-Thomas, Y., Bridgewater, P., Tengo, M., Thaman, R., Yao, C., Berkes, F., Carino, J., Carneiro da Cunha, M., Diaw, M., Diaz, S., Figueroa, V., Fisher, J., Hardison, P., Ichikawa, K., Kariuki, P., Karki, M., Lyver, P., Malmer, P., Masardule, O., Yeboah, A., Pacheco, D., Pataridze, T., Perez, E.,

Roue, M., Roba, H., Rubis, J., Saito, O., & Xue, D. (2020). Working with indigenous, local, and scientific knowledge in assessments of nature and nature's linkages with people. *Current Opinion in Environmental Sustainability*, 43, 8–20.

Howson, P. (2020). Building trust and equity in marine conservation and fisheries supply chain management with blockchain. *Marine Policy*, 115, 103873.

Hussain, N., Rigoni, U., & Orij, R. P. (2019). The role of corporate sustainability performance in organizational attractiveness: A behavioral reasoning perspective. *Journal of Business Ethics*, 154(2), 395–409.

Iansiti, M., & Lakhani, K. R. (2017). The truth about blockchain. *Harvard Business Review*, 95(1), 118–127.

IBM. (2021). *IBM Food Trust*. IBM. Retrieved from: https://www.ibm.com/products/supply-chain-intelligence-suite/food-trust.

Intel. (2019). *Secure Your Business: End-to-end Supply Chain Traceability*. Intel. Retrieved from: https://tsc.intel.com/documents/TSCBlockchain_white_paperFINAL.PDF.

Intergovernmental Panel on Climate Change. (2014). *Climate Change 2014: Synthesis Report*. Contribution of Working Groups I, II and III to the Fifth Assessment Report of the Intergovernmental Panel on Climate Change [Core Writing Team, R. K. Pachauri, & L. A. Meyer (Eds.)]. IPCC.

International Telecommunication Union. (2020). *Artificial Intelligence for Sustainable Development, Report*. International Telecommunication Union.

Johnson, L., & Turner, P. (2023). *Integrating Emerging Technologies into Higher Education*. Academic Press.

Johnson, R. (2019). Enhancing third-party verification in sustainability reporting through blockchain. *Corporate Social Responsibility and Environmental Management*, 26(2), 400–408.

Kamath, R. (2018). Food traceability on blockchain: Walmart's pork and mango pilots with IBM. *The Journal of the British Blockchain Association*, 1(1).

Kent, S. J., Jones, G. A., Zhambyl, S., & Kappen, J. A. (2024). Communicating sustainability through language differences with rich point pedagogy. In *Handbook of Social Sustainability in Business and Management*. Springer.

King, S., & Nadal, S. (2012). *PPCoin: Peer-to-Peer Crypto-Currency with Proof-of-Stake*. Self-Published White Paper.

KPMG International. (2020). *The KPMG Survey of Corporate Responsibility Reporting 2020*. KPMG International.

Krause, M. J., & Tolaymat, T. (2018). Quantification of energy and carbon costs for mining cryptocurrencies. *Nature Sustainability*, 1(11), 711–718.

Kshetri, N. (2022). Blockchain systems and ethical sourcing in the mineral and metal industry: A multiple case study. *The International Journal of Logistics Management*, 33(1), 1–27.

Larimer, D. (2014). *Delegated Proof-of-Stake (DPOS)*. Bitshares Whitepaper.

Lee, M., & Nguyen, H. (2022). Partnerships in technology for sustainable development. *Journal of Sustainable Innovation*, 11(2), 45–59.

Li, S., & Badinelli, R. (2022). Blockchain-promoted transparency in sustainability reporting: A literature review. In R. Badinelli & S. Li (Eds.), *Handbook of Blockchain and Sustainable Development* (pp. 1–20). Springer.

Lozano, R., & Huisingh, D. (2011). Inter-linking issues and dimensions in sustainability reporting. *Journal of Cleaner Production*, 19(2–3), 99–107.

LVMH. (2021). *LVMH Partners with Other Major Luxury Companies on Aura, the First Global Luxury Blockchain*. LVMH.

Manetti, G. (2011). The quality of stakeholder engagement in sustainability reporting: Empirical evidence and critical points. *Corporate Social Responsibility and Environmental Management*, 18(2), 110–122.

Manetti, G., & Becatti, L. (2009). Assurance services for sustainability reports: Standards and empirical evidence. *Journal of Business Ethics*, 87(1), 289–298.

Manetti, G., & Toccafondi, S. (2012). The role of stakeholders in sustainability reporting assurance. *Journal of Business Ethics*, 107(3), 363–377.

Meadows, J., Annandale, M., & Ota, L. (2019). Indigenous peoples' participation in sustainability standards for extractives. *Land Use Policy*, 88, 104118.

Mejbel, A. K., & Salman, A. M. (2024). The impact of integrated reporting on company continuity. *International Journal of Business and Management Studies*, 4(3), 8–32.

Michelon, G., Pilonato, S., & Ricceri, F. (2015). CSR reporting practices and the quality of disclosure: An empirical analysis. *Critical Perspectives on Accounting*, 33, 59–78.

Miglani, A., Kumar, N., Chamola, V., & Zeadally, S. (2020). Blockchain for internet of energy management: Review, solutions, and challenges. *Computer Communications*, 151, 395–418.

Mougayar, W. (2016). *The Business Blockchain: Promise, Practice, and Application of the Next Internet Technology.* John Wiley & Sons.

Mukhamedova, Z., & Mukhamedova, D. (2023). Prospects of using blockchain technology in the organization of the transportation process and supply chain. *International Journal of Intelligent Systems and Applications in Engineering*, 12(2s), 379–387.

Naheed, R., Waqas, M., Ahmad, N., & Iqbal, M. (2024). Fostering sustainability and green innovation reporting in manufacturing firms: An investigation of barriers through ISM-MICMAC approach. *Environment, Development and Sustainability*, 1–43.

Nakomoto, S. (2008). *A Peer-to-peer Electronic Cash System.* White Paper.

Nidumolu, R., Prahalad, C. K., & Rangaswami, M. R. (2009). Why sustainability is now the key driver of innovation. *Harvard Business Review*, 87(9), 56–64.

O'Dwyer, B., Owen, D. L., & Unerman, J. (2011). Seeking legitimacy for new assurance forms: The case of assurance on sustainability reporting. *Accounting, Organizations and Society*, 36(1), 31–52.

O'Faircheallaigh, C. (2020). Impact and benefit agreements as monitoring instruments in the minerals and energy industries. *The Extractive Industries and Society*, 7(4), 1338–1346.

Ostrowski, W. (2020). Transparency and global resources: Exploring linkages and boundaries. *The Extractive Industries and Society*, 7(4), 1472–1479.

Patel, S. (2022). Cultural development and sustainability in corporate settings. *Journal of Corporate Citizenship*, (68), 34–52.

Pilkington, M. (2016). Blockchain technology: Principles and applications. In F. Xavier Olleros & M. Zhegu (Eds.), *Research Handbook on Digital Transformations*. Edward Elgar Publishing.

Porter, M. E., & Kramer, M. R. (2011). Creating shared value. *Harvard Business Review*, 89(1/2), 62–77.

Prendergast, J., & Lezhnev, S. (2009). From mine to mobile phone. In *The Conflict Minerals Supply Chain*. Washington, DC: Enough Project.

Provenance. (2021). *Provenance, Issue 19, 2021 ISSN: 1832-2522.*

Read, I. I., & Pehlivan, C. N. (2020). Blockchain and data protection: A compatible couple? *Global Privacy Law Review*, 1(1).

Schaltegger, S., & Burritt, R. (2006). Corporate sustainability accounting: What is it and what is it good for? *Australian Accounting Review*, 16(3), 67–73.

Schaltegger, S., Burritt, R., & Petersen, H. (2016). *An Introduction to Corporate Environmental Management: Striving for Sustainability.* Springer Science & Business Media.

Shen, C., & Pena-Mora, F. (2018). Blockchain for cities-a systematic literature review. *IEEE Access*, 6, 76787–76819.

Shishegaran, A., Safari, S., & Karami, B. (2022). Sustainability evaluation for selecting the best optimized structural designs of a tall building. *Sustainable Materials and Technologies*, 33, e00482.

Smith, J., & Lee, K. (2020). Blockchain in sustainability reporting: A game changer. *Sustainability*, 12(15), 6154.

Sosa, I., & Keenan, K. (2001). *Impact Benefit Agreements between Aboriginal Communities and Mining Companies: Their Use in Canada.* Report: Canadian Environmental Law Association, Environmental Mining Council of British Columbia, and Cooper Accion: Accion Solidaria para el Desarrollo.

Stach, C., Gritti, C., Przytarski, D., & Mitschang, B. (2022, April). Can blockchains and data privacy laws be reconciled? A fundamental study of how privacy-aware blockchains are feasible. In *Proceedings of the 37th ACM/SIGAPP Symposium on Applied Computing* (pp. 1218–1227).

Supply Chain Magazine. (2020). *UPS Joins Alliance to Create Blockchain Standards for Logistics, Report.* Supply Chain Magazine.

Svensson, G., & Wagner, B. (2021). Transparent reporting as a byproduct of capacity development: A comparative study. *Journal of Cleaner Production*, 285, Article 125293.

Swan, M. (2015). *Blockchain: Blueprint for a New Economy.* O'Reilly Media.

Trans Mountain Corporation. (2023). *Trans Mountain: 2023 Environmental, Social & Governance Report.* Retrieved July, 2024 from: https://docs.transmountain.com/ESG-Reports/TransMountain_2023-ESG-Report.pdf#asset:502271:url.

United Nations Environment Programme (UNEP). (2019). *The Emissions Gap Report 2019.* United Nations Environment Programme (UNEP).

United Nations Environment Programme (UNEP). (2021). *Blockchain and DLT for Sustainability.* Retrieved from: www.unep.org/resources/report/blockchain-and-dlt-sustainability.

United Nations Environment Programme (UNEP). (2022). *Blockchain for Sustainable Energy and Climate in the Global South.* United Nations Environment Programme (UNEP).

United Nations Framework Convention on Climate Change. (2015). *Paris Agreement, Convention Paris in 2015.* United Nations.

United Nations. (2008). *United Nations Declaration on the Rights of Indigenous Peoples.* United Nations.

United Nations. (2015). *Transforming Our World: The 2030 Agenda for Sustainable Development.* United Nations.

VeChain Foundation. (2021). *VeChain ToolChain: Revolutionizing Business with the Blockchain (Report).* VeChain Foundation.

Vigna, P., & Casey, M. J. (2015). *The Age of Cryptocurrency: How Bitcoin and Digital Money are Challenging the Global Economic.*

World Commission on Environment and Development. (1987). *Our Common Future (The Brundtland Report).* Oxford University Press.

World Resources Institute, & World Business Council for Sustainable Development. (2011). *Greenhouse Gas Protocol: Corporate Value Chain (Scope 3) Accounting and Reporting Standard.* World Resources Institute and World Business Council for Sustainable Development.

World Wide Fund for Nature. (2023). *Blockchain Innovations in Environmental Protection (Report).* World Wide Fund for Nature.

9 Blockchain for Carbon Credits and Emissions Reduction

Mojtaba Rezaie and Aydin Shishegaran

9.1 INTRODUCTION

Climate change is an undeniable threat (Onoh et al., 2024), and reducing carbon emissions is a global priority (Akpuokwe et al., 2024). Carbon credits and tradable units representing verified emission reductions have offered a market-based solution for making a healthy global environment. However, the traditional carbon markets face challenges like fraud, double counting, and lack of transparency (Danish et al., 2024). This is where blockchain technology is welcomed as a game changer. Because of its safe and secure nature, it can play a prominent role in revolutionizing carbon credit systems (Vilkov & Tian, 2023). Due to blockchain capabilities, the assumption is that this technology promises to promote trust, efficiency, and, ultimately, a more sustainable future in the carbon credits market.

According to Mapped, the changes in global greenhouse gas emissions measures in tons of carbon dioxide-equivalents (CO_2eq) over a 100-year timescale, significant growth was indicated from 2000 to 2022. This means that the rate of CO_2eq in 2000 was 40.77 billion tons, and its growth will be 53.85 in 2022 (Jones et al., 2023) due to Our World in Data. Global warming and climate challenges remain if all the carbon dioxide, methane, and nitrous oxide sources continue to be realized through land-use change. This problem is now known as a new-century climate change challenge (Ritchie et al., 2020). Kyoto Protocol (Böhringer, 2003) was established to regulate states to cooperate in reducing carbon emissions as the 1992 United Nations Framework Convention on climate change. It was considered the carbon trading system. Recently, the Paris Agreement under the enhanced transparency framework (ETF) considers an international treaty on climate change as a report transparently on actions taken and progress in climate change mitigation, as well as adaptation measures and support provided or received (Weikmans et al., 2021).

9.2 CARBON CREDIT MARKET

Carbon credits are tradable units that allow the holder to emit a certain amount of greenhouse gas (GHG), typically one ton of carbon dioxide equivalent (Kousika & Vennila, 2023). These credits are present in two forms: (1) permits and (2)

DOI: 10.1201/9781003609865-9

certificates. Permits grant permission to emit a specific amount, while certificates verify that an equivalent amount of emissions has been avoided or removed elsewhere.

The Kyoto Protocol introduced the Emissions Trading System (ETS) to promote international cooperation on climate change through credit trading (Show & Lee, 2008). This system is a practical policy for participants with surplus credits to sell to others who need them to meet government-mandated emission limits. As a result, many countries have adopted the ETS framework to manage their carbon emissions (Kenton, 2023).

The experiences in this field indicated that currently, in global environmental conditions due to greenhouse gas emissions, the focus must be on the uses of renewable energies. For example, the US policies target to drive its power of grid by 2035 to 100% renewables, and ambitious goals like the Paris Agreement's 85% emission reduction target for the building sector (Ahmed et al., 2020) have been highlighted global directions for reducing emissions and making a sustainable future for the next generation of humans. However, achieving such ambitious goals requires robust monitoring, reporting, and verification (MRV) systems. Without a proper MRV system, the risk of implementing such plans remains high (Woo et al., 2021).

In the building sector, participation in the ETS incentivizes lower emissions. Stakeholders can benefit from their self-MRV raises by trading any excess credits generated through the carbon credit market. This system creates market-based pressure to reduce emissions while rewarding those who can demonstrate their efforts through effective monitoring (Woo et al., 2021).

Carbon credit markets are grouped into three major categories: (1) mandatory versus voluntary, (2) allocation versus offset, and (3) international versus regional markets, as shown in Figure 9.1 (Woo et al., 2020).

The previous figure indicates that the mandatory market has been regulated by law or a mandatory reduction target of greenhouse gases (GHGs), and the voluntary market is based upon the cooperation of individuals. Allocation markets focus on trade emission allowances that the government allocates to industrial facilities, fuel suppliers, and electricity importers. Moreover, in contrast, offset markets provide an opportunity for companies or projects to invest in carbon reduction aimed to give climate action projects to compensate for their GHG emissions; afterward, International markets allow carbon credit trading between countries, while regional markets restrict transactions within a country's borders.

ETS typically covers manufacturing, energy production, power generation, transportation, and waste management. Only Japan allows the building sector to participate in carbon credit trading (Fund & Solutions, 2016; Partnership, 2021; Zhang, 2020). The one key element for controlling global warming is the nationally determined contributions (NDCs) made by countries that pledge specific actions to reduce emissions. This action is now encompassing the building sector. However, it could have been an efficient activity among commitment countries. Because 136 countries included the building sector in their NDCs by 2018, most still do not need to drive their policies for this sector aimed at cutting emissions (Construction, 2018).

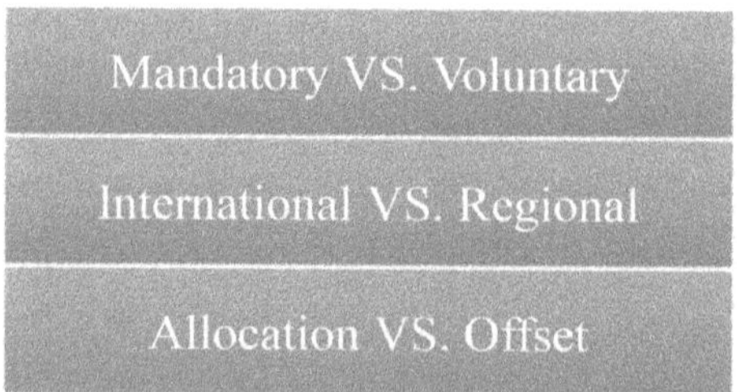

FIGURE 9.1   Types of carbon credit markets.

9.3   ADOPTION OF BLOCKCHAIN TECHNOLOGY IN THE CARBON CREDIT MARKET

Blockchain is an innovative emerging technology in the carbon credit market (Parhamfar et al., 2024), and it also holds immense potential to transform this type of market. By enhancing transparency (Spilker & Nugent, 2022), streamlining transactions, and improving efficiency (Yang et al., 2023), blockchain can contribute significantly to a more effective and impactful carbon credit system. Rather, this technology is used as a powerful tool that drives the benefits of the digital transformation approach in ETS; due to its potential application, blockchain can revolutionize the carbon credit market. Here is how blockchain is transforming this space.

Enhanced Transparency and Trust. Traditional carbon credit markets are challengeable due to their concerns about double counting and the true impact of credits. Blockchain provides a secure and transparent ledger system where every carbon credit transaction is recorded immutably. This allows all stakeholders to track each credit's origin, ownership, and expiration, fostering trust and confidence in the market (Rana et al., 2023).

Streamlined Transactions and Reduced Costs. Blockchain facilitates peer-to-peer trading, enabling direct transactions between buyers and sellers. This cuts out intermediaries, simplifies the process, and potentially reduces transaction fees (Blanton et al., 2024).

Improved Efficiency and Accessibility. Complex verification processes can hinder participation in the carbon credit market. So, blockchain's smart contracts can automate key steps, streamlining verification and making the market more accessible to smaller players who might have previously been excluded (Parhamfar et al., 2024).

There are many initiatives for applying blockchain usage in carbon credits. Platforms are being developed to track credits from renewable energy projects (Yuan et al., 2023), forestry initiatives (Blanton et al., 2024), and other emission reduction efforts. With the adoption of technology in this market, it matures, and regulations adapt; it is realistic to expect future innovative applications to emerge. Tracking renewable energy projects like Chjango and Australian Power Ledger, Verra Registry and IM Forestry Initiatives, or WEForest and AirCarbon, as well as others, are proven examples of the adoption of blockchain in the carbon credit market.

9.3.1 The Role of Carbon Credits in a Sustainable Building Sector

A 2013 study by Zaid (Zaid & Suzaini, 2013) found that the construction industry is a major contributor to global energy use and carbon emissions, accounting for roughly 40%. This sector is responsible for consuming nearly 60% of the world's power, with a staggering 80–90% of that energy used during the operational phase of buildings. Based on breaking down the emissions further, residential buildings are responsible for two-thirds of the global building sector's carbon footprint, while commercial buildings contribute the remaining one-third. Importantly, like energy consumption, 80–90% of these emissions occur during the operation of buildings.

The previous figure indicates that the building sector has significant potential for efficiency gains in control of climate change globally, energy consumption reduction, and CO_2 emissions reductions. The figure indicates a maximum achievable reduction of 75% in energy use and 35% in emissions related to building operations. This highlights the substantial opportunity for positive change within this sector.

The 2019 Green Globes Assessment Protocol focuses on reducing carbon dioxide equivalent (CO_{2e}) emissions. This protocol outlines a specific reduction scheme where a building achieves a 50% decrease in CO_2e emissions compared to a baseline building in its region. This reduction is calculated using the following formula, which is defined as a reduction in the presence of CO_2e:

$$CO_{2e} = 100X\left(1 - PER/BER\right) \tag{1}$$

where Baseline Equivalent Emission Rate (BER) represents the standard CO_2e emission rate for a typical building in the same region as the proposed building, and the Proposed Building's CO_2e Emission Rate (PER) is the actual CO_2e emission rate of the building being assessed (ANSI/GBI, 2019). This protocol is a valuable tool for promoting greener commercial buildings, and by providing a comprehensive framework and encouraging energy efficiency, the protocol contributes to a more sustainable built environment.

The limited impact of individual buildings, overwhelming efficiency options, misaligned motivations, and insufficient incentives address climate change challenges in the building sector. The ETS is used as a solution to overcome these challenges. It saves buildings energy, simplifies MRV monitoring, and helps building owners and occupants adopt energy-efficient practices (Fernando et al., 2021).

9.3.2 Blockchain and Carbon Trading

Manufacturing companies face growing pressure to minimize greenhouse gas (GHG) emissions during production as environmental concerns rise. Blockchain technology as a platform has emerged for emissions trading and targets to minimize GHG (Khaqqi et al., 2018). This system, called an Emissions Trading Scheme (ETS), essentially functions as the tradable permit policy. A set number of permits allowing a specific amount of GHG emissions are distributed to participating companies. At the end of a designated period, companies report their actual emissions. Companies that have produced less pollution than their permits allow can then sell their excess permits to those who have exceeded their limit. This creates an economic incentive for manufacturers to minimize emissions. More efficient companies can profit by selling unused permits, while those exceeding their limits face the cost of purchasing additional permits (Fernando et al., 2021).

Emission Trading Schemes (ETS) hold promise for manufacturers looking to reduce their carbon footprint. However, traditional methods can be cumbersome and need more transparency. Blockchain technology offers a solution that enhances these aspects significantly. First, blockchain creates a secure and transparent shared ledger for recording all emission trading transactions (Saberi et al., 2019). This ensures the legitimacy and accuracy of each trade, preventing errors or manipulation. Every transaction is timestamped and permanently recorded, allowing for complete traceability and eliminating the possibility of lost quotas or duplicate transactions. Additionally, the system requires all transactions to adhere to the same consensus mechanism, guaranteeing consistency across the platform (Pan et al., 2019). This eliminates confusion and inconsistencies that can plague traditional systems.

9.3.3 BLOCKCHAIN, MANUFACTURING SECTOR, AND ENERGY CONSUMPTION

Blockchain technology offers exciting possibilities for improving efficiency in energy management and control. By utilizing mining algorithms, blockchain can analyze energy consumption patterns and identify areas for sustainability improvements. Additionally, integrating artificial intelligence (AI) and virtual energy management systems allows for optimizing energy use, such as pinpointing the optimal placement of heat recovery systems within the blockchain-distributed ledger. The Internet of Things (IoT) can further enhance the productivity of such systems by providing real-time monitoring of energy consumption (Bürer et al., 2019). This empowers authorized personnel with access to transparent and reliable data on energy use, which can be visualized to differentiate between renewable and non-renewable energy sources (Teufel et al., 2019).

However, it is important to consider the broader sustainability implications of blockchain technology, from both behavioral and socioeconomic perspectives (Bürer et al., 2019). Technically, challenges remain in ensuring grid stability while monitoring energy usage within blockchain systems (Parnell, 2017). While some argue that Bitcoin, a prominent blockchain platform, effectively tracks and visualizes energy consumption patterns (Bitcoinmining.com, 2017), concerns exist regarding its scalability and long-term sustainability (de Vries, 2018), and maybe this can be expanded to other blockchain environmental functions. Ultimately, the success of blockchain technology in reducing energy consumption hinges on its ability to minimize the environmental impact caused by fossil fuel use. If implemented effectively, blockchain can promote the development and adoption of renewable energy sources, ultimately contributing to global carbon reduction efforts (Wang & Su, 2020).

Carbon dioxide (CO_2) is the most prevalent greenhouse gas generated by human activities, making monitoring critical for tracking progress toward emission reduction targets (Pachauri et al., 2014). The manufacturing sector is known as the second-largest contributor to CO_2 emissions, and firms in this sector should focus on reducing the carbon emissions of products by using energy efficiency and applying policies related to tradable carbon credits. Therefore, manufacturers must prioritize reducing the carbon footprint of their products by focusing on energy efficiency (Böttcher & Müller, 2015) and regulatory carbon credits.

Composite indicators offer a valuable tool for monitoring low-carbon performance within the manufacturing sector (Zhang & Zhou, 2018). These indicators can synthesize complex and multifaceted data points into a single, clear metric. Governments can play a crucial role in promoting sustainability by educating firms about clean production practices and offering incentives to reduce their carbon footprints (Böttcher & Müller, 2015). By implementing these strategies, the manufacturing sector can significantly contribute to global efforts to mitigate climate change.

9.4 BLOCKCHAIN AND CARBON CREDIT MARKET

Blockchain technology transforms the carbon credit market by promoting transparency, efficiency, and accessibility. By evidence of the impact of the carbon credit market on the reduction of greenhouse emissions, it is inferred that the relationship between blockchain technology and the carbon credit market is defined as an action for directing the unleashed potential of the green economy and providing a healthy environmental ecosystem.

Environmental groups use blockchain, a secure online ledger system, to stabilize carbon credit markets. These credits, offered by organizations like Veridium and Nori, represent verified emissions reductions. This approach has created a new "Blockchain carbon market". The long-term vision is a global network where carbon credits can be easily traded like digital assets. This would ensure accurate tracking of emissions while benefiting owners of eco-friendly buildings. Simplifying the process could lead to wider public backing for green initiatives (Woo et al., 2020).

The following content indicates the significant potential of blockchain technology in the carbon credit market. It also explains how blockchain can be implemented and the positive impact it is expected to have on the market as a whole.

The carbon credit market mitigates climate change. Organizations can offset emissions by investing in projects that reduce greenhouse gases elsewhere. However, traditional carbon credit markets have faced challenges related to transparency, double counting, and accessibility. With its distributed ledger system and tamper-proof records, blockchain technology offers a promising solution to address these limitations.

One of the key benefits of blockchain in the carbon credit market is its ability to enhance transparency and traceability. Carbon credits are represented as tokens on the blockchain, providing an immutable record of their issuance, ownership, and retirement. This allows all participants in the market to track the origin and validity of credits, reducing the risk of fraud and double counting. Additionally, smart contracts can be programmed to automate verification, ensuring that credits meet predetermined environmental standards.

Blockchain facilitates peer-to-peer transactions between buyers and sellers of carbon credits, eliminating the need for intermediaries. This can significantly reduce transaction costs associated with verification, registration, and brokerage fees. Additionally, blockchain platforms can enable fractional ownership of carbon credits, making them more accessible to smaller investors and project developers. This increased liquidity could stimulate market activity and drive down the overall cost of carbon credits.

Blockchain technology can improve the efficiency and integration of carbon credit markets globally. Standardized protocols can be established on blockchain platforms to ensure consistency in credit issuance and verification across different jurisdictions. This can facilitate cross-border trading and encourage the development of a more unified global carbon market. Furthermore, blockchain-based carbon credit platforms can integrate with existing environmental data sources, enabling real-time monitoring of project performance and the environmental impact of carbon offsets.

These practical potentials and benefits are written due to the challenges of adopting blockchain technology in the carbon credit market. Standardizing the blockchain across different platforms is essential to ensure interoperability and prevent market fragmentation. Additionally, ensuring the underlying blockchain technology's energy efficiency is crucial to maintaining its environmental sustainability. Regulatory frameworks need to adapt to address issues related to tokenization and smart contracts in the context of carbon credits.

9.5 BLOCKCHAIN APPLICATION IN CARBON TRADING SYSTEMS

Blockchain is a technology based on distributed ledgers. This instrument is a database that all members of a network share. This means that all participants (members) in the network have access to the same information in the database, and this can be accomplished even from remote or multiple sites if the user knows their code. Because all parties can possess an identical copy of the ledger, any modifications are updated in all copies in a short time. The data can be financial, contractual, electronic, and so forth. The information is stored cryptographically by the key signatures (Government, 2016). This guarantees security and accuracy. The cyber-sheets (copies of the database stored on individual computers) are carried over decentralized "nodes" within networks. All users trust node storage because verification lies in common knowledge. This veracity is provided in the computational code (Li et al., 2019). These concepts can be the backbone of interconnective security (Woo et al., 2020).

Blockchain's introduction focused on cryptocurrency but has recently pushed toward further horizons. According to the results of unfettered transactions and circulation of a panoply of assets, it is an innovation. This technology has created a transparent and secure product using a consensus algorithm. The blockchain paradigm organizes processes with less cost and more efficiency because an intermediary does not need oversight. Blockchain can operate at a scale that, heretofore, was impossible. This method aims to automate resource allocation for physical and non-physical assets (Swan, 2015), which can be tradable digitally.

Blockchain falls into three categories: (1) public, (2) private, and (3) consortium (permission) realms. The public option possesses no limitation. Anyone may access it. Private networks give that access right only to certified members in advance.

A consortium blockchain combines public and private attributes, but permission is needed.

Blockchain networks are based on a core principle called authorized participation. Members must prove their legitimacy before conducting transactions. This ensures that everyone on the network is who they say they are. Decision-making is another key feature. All participants must reach a consensus agreement before adding a new block to the chain. This ensures everyone has the same, unalterable copy of the information. Once an agreement is reached, the network works together to create and secure these blocks using cryptographic chains (hashes). There are various methods for achieving consensus, each with its advantages.

> *Proof of Work (PoW).* This is the original method that Bitcoin uses. It requires solving complex mathematical problems to validate transactions.
> *Proof of Stake (PoS).* This method relies on members holding a stake in the network (cryptocurrency) to validate transactions. Practical Byzantine Fault Tolerance (PBFT).
> *Delegated Proof of Stake (DPOS).* These are more specialized consensus mechanisms used in specific blockchain applications.

Smart contracts are a powerful feature of blockchain technology. These are self-executing contracts written in code. When predefined conditions are met, the code automatically executes the terms of the agreement, eliminating the need for intermediaries. These contracts can be programmed to handle anything from bureaucratic procedures to legal regulations using various programming languages like Java, Go, and Python (Nawari & Ravindran, 2019).

Hyperledger, an open-source platform backed by the Linux Foundation, allows for creating customized blockchain solutions. For instance, Hyperledger Fabric, known for its modular design, facilitates secure and efficient transactions. It also reduces the computational burden of proof of work (PoW) (Gupta, 2017). This platform was instrumental in Veridium and IBM's collaboration to create a blockchain-powered carbon credit market (Babich & Hilary, 2019).

Tokenization is another significant concept. It involves converting legal ownership of an asset into a digital token. This streamlines transactions for traditionally less liquid assets and facilitates micropayments. Veridium and IBM's carbon credit market exemplify how tokenization can represent emission levels (Babich & Hilary, 2019).

The construction industry is also exploring the potential of blockchain. Companies like Amsterdam-based HerenBouw have implemented blockchain-based project management systems to boost efficiency. Similarly, Californian blockchain company Briq demonstrates how distributed ledgers can revolutionize document storage and sharing within construction projects (Vargas, 2019). These examples showcase the immense potential of blockchain technology in transforming the construction industry.

As blockchain technology matures and regulations adapt, these practical applications can be expected to expand and contribute to a more efficient and impactful global carbon market while blockchain technology is still evolving in the carbon trading space. The following are examples of pilot projects and initiatives related to blockchain potential in carbon credit systems.

Veridium and IBM's GreenMark Platform. This collaboration leverages Hyperledger Fabric, an open-source blockchain platform, to create a transparent and secure marketplace for carbon credits. Veridium, a leading environmental solutions provider, issues tokenized carbon credits based on verified emission reduction projects. These tokens represent ownership of the emissions reductions and can be traded on the GreenMark platform. IBM's blockchain expertise ensures secure transactions and streamlined record-keeping. This project fosters trust and transparency in the carbon credit market, allowing buyers to be confident in the environmental impact of their purchases.

China's National Carbon Emissions Trading Scheme (NCETS) Pilot. China, the world's largest emitter, has embarked on ambitious pilot programs using blockchain technology for its national carbon trading scheme. Those activities relate to issuing and tracking emissions allowances, streamlined compliance reporting, and enhanced market efficiency.

The AirCarbon Project. This initiative focuses on the voluntary carbon market and utilizes blockchain technology to connect project developers with businesses and individuals seeking to offset their carbon footprint. The project uses a transparent system for creating and tracking carbon offsets associated with various initiatives like renewable energy projects and forest conservation efforts. Individuals and businesses can purchase verified carbon offset tokens on the platform, supporting these projects and contributing to positive environmental change.

9.6 SUMMARY

This chapter explores how blockchain technology can transform the carbon credit market, particularly within the building sector, by making it more transparent and efficient. Carbon credit markets encourage organizations to reduce emissions by allowing them to offset their emissions through verified carbon credit projects. Blockchain's distributed ledger mechanism ensures secure, transparent, and traceable transactions, which can enhance the reliability of carbon trading systems.

The chapter highlights several initiatives demonstrating blockchain's potential to reduce carbon emissions, including the development of open-source blockchain platforms, government pilot plans, and projects like AirCarbon. For blockchain to be effectively integrated into the carbon credit market, the authors recommend standardizing the system through a globally governed multi-platform, fostering collaboration among leading environmental protection organizations.

REFERENCES

Ahmed Ali, Khozema, Mardiana I. Ahmad, and Yusri Yusup. 2020. "Issues, Impacts, and Mitigations of Carbon Dioxide Emissions in the Building Sector." *Sustainability* 12(18).

Akpuokwe, Chidiogo Uzoamaka, Adekunle Oyeyemi Adeniyi, Seun Solomon Bakare, and Nkechi Emmanuella Eneh. 2024. "Legislative Responses to Climate Change: A Global Review of Policies and Their Effectiveness." *International Journal of Applied Research in Social Sciences* 6(3):225–239.

ANSI/GBI. 2019. ANSI/GBI 01–2019 Green Globes Assessment Protocol for Commercial Buildings.

Babich, V., and G. Hilary. (2019). "Distributed Ledgers and Operations: What Operations Management Researchers Should Know about Blockchain Technology to Cite This Version: HAL Id: Hal-02005158 Distributed Ledgers and Operations: What Operations Management Researchers Should Know about." Manufacturing and Service Operations Management, INFORMS, In Press, Hal 20051.

Bitcoinmining.com. "Bitcoin Mining Hardware Guide." 2017.

Blanton, Austin, Midhun Mohan, G. A. Pabodha Galgamuwa, Michael S. Watt, Jorge F. Montenegro, Freddie Mills, Sheena Camilla Hirose Carlsen, Luisa Velasquez-Camacho, Barbara Bomfim, Judith Pons, Eben North Broadbent, Ashpreet Kaur, Seyide Direk, Sergio de-Miguel, Macarena Ortega, Meshal Abdullah, Marcela Rondon, Wan Shafrina Wan Mohd Jaafar, Carlos Alberto Silva, Adrian Cardil, Willie Doaemo, and Ewane Basil Ewane. 2024. "The Status of Forest Carbon Markets in Latin America." *Journal of Environmental Management* 352:119921.

Böhringer, Christoph. 2003. "The Kyoto Protocol: A Review and Perspectives." *Oxford Review of Economic Policy* 19(3):451–466.

Böttcher, Christian Felix, and Martin Müller. 2015. "Drivers, Practices and Outcomes of Low-Carbon Operations: Approaches of German Automotive Suppliers to Cutting Carbon Emissions." *Business Strategy and the Environment* 24(6):477–498.

Bürer, Mary Jean, Matthieu de Lapparent, Vincenzo Pallotta, Massimiliano Capezzali, and Mauro Carpita. 2019. "Use Cases for Blockchain in the Energy Industry Opportunities of Emerging Business Models and Related Risks." *Computers & Industrial Engineering* 137:106002.

Danish, Syed Muhammad, Kaiwen Zhang, Fatima Amara, Juan Carlos Oviedo Cepeda, Luis Fernando Rueda Vasquez, and Tom Marynowski. 2024. "Blockchain for Energy Credits and Certificates: A Comprehensive Review." *IEEE Transactions on Sustainable Computing* (01):1–13.

de Vries, Alex. 2018. "Bitcoin's Growing Energy Problem." *Joule* 2(5):801–805.

Fernando, Yudi, Nor Hazwani Mohd Rozuar, and Fineke Mergeresa. 2021. "The Blockchain-Enabled Technology and Carbon Performance: Insights from Early Adopters." *Technology in Society* 64:101507.

Fund, Environmental Defense, and Climate Challenges Market Solutions. "Republic of Korea: An Emissions Trading Case Study." 2016. https://www.edf.org/sites/default/files/korean_case_study.pdf.

Government, UK. 2016. "Distributed Ledger Technology: Beyond Blockchain." Crown.

Gupta, Manav. 2017. *Blockchain for Dummies (IBM Limited Edition)*. Hoboken.

Jones, Matthew W., Glen P. Peters, Thomas Gasser, Robbie M. Andrew, Clemens Schwingshackl, Johannes Gütschow, Richard A. Houghton, Pierre Friedlingstein, Julia Pongratz, and Corinne Le Quéré. 2023. "National Contributions to Climate Change Due to Historical Emissions of Carbon Dioxide, Methane, and Nitrous Oxide since 1850." *Scientific Data* 10(1):155.

Kenton, W. "Carbon Credits and How They Can Offset Your Carbon Footprint." *Investopedia*, 2023.

Khaqqi, Khamila Nurul, Janusz J. Sikorski, Kunn Hadinoto, and Markus Kraft. 2018. "Incorporating Seller/Buyer Reputation-Based System in Blockchain-Enabled Emission Trading Application." *Applied Energy* 209:8–19.

Kousika, K., and Dr. D. Vennila. June 2023. "Carbon Credit and Trading: An Overview." *International Journal of Innovative Science and Research Technology (IJISRT)* 8(6): 180–183. www.ijisrt.com. ISSN – 2456-2165. https://doi.org/10.5281/zenodo.8047379.

Li, Jennifer, David Greenwood, and Mohamad Kassem. 2019. "Blockchain in the Construction Sector: A Socio-Technical Systems Framework for the Construction Industry." Pp. 51–57 in *Advances in Informatics and Computing in Civil and Construction Engineering: Proceedings of the 35th CIB W78 2018 Conference: IT in Design, Construction, and Management.*

Nawari, Nawari O., and Shriraam Ravindran. 2019. "Blockchain and Building Information Modeling (BIM): Review and Applications in Post-Disaster Recovery." *Buildings* 9(6).

Onoh, Udochukwu C., Josiah Ogunade, Emmanuel Owoeye, Solomon Awakessien, and Joseph Kwesi Asomah. 2024. "Impact of Climate Change on Biodiversity and Ecosystems Services." *International Journal of Geography and Environmental Management (IJGEM)* 10(1):77–93.

Pachauri, Rajendra K., Myles R. Allen, Vicente R. Barros, John Broome, Wolfgang Cramer, Renate Christ, John A. Church, Leon Clarke, Qin Dahe, Purnamita Dasgupta, and others. 2014. "Climate Change 2014: Synthesis Report." Contribution of Working Groups I, II and III to the Fifth Assessment Report of the Intergovernmental Panel on Climate Change. Ipcc.

Pan, Yuting, Xiaosong Zhang, Yi Wang, Junhui Yan, Shuonv Zhou, Guanghua Li, and Jiexiong Bao. 2019. "Application of Blockchain in Carbon Trading." *Energy Procedia* 158:4286–4291.

Parhamfar, Mohammad, Iman Sadeghkhani, and Amir Mohammad Adeli. 2024. "Towards the Net Zero Carbon Future: A Review of Blockchain-Enabled Peer-to-Peer Carbon Trading." *Energy Science & Engineering* 12(3):1242–1264.

Parnell, John. 2017. "The Sun Exchange: Bitcoin Is the Answer to Africa's Solar Finance Headaches." PVtech.

Rana, Muhammad Tayyab, Muhammad Numan, Muhammad Yousif, Tanveer Hussain, and Akif Zia Khan. 2023. "Blockchain Technology for Electric Vehicles: Applications, Challenges, and Opportunities in Charging Infrastructure, Energy Trading, Energy Management, and Supply Chain." Challenges and Opportunities in Charging Infrastructure, Energy Trading, Energy Management, and Supply Chain.

Ritchie, Hannah, Pablo Rosado, and Max Roser. 2020. "Greenhouse Gas Emissions." Our World in Data.

Saberi, Sara, Mahtab Kouhizadeh, Joseph Sarkis, and Lejia Shen. 2019. "Blockchain Technology and Its Relationships to Sustainable Supply Chain Management." *International Journal of Production Research* 57(7):2117–2135.

Show, K. Y., and Duu-Jong Lee. 2008. "Carbon Credit and Emission Trading: Anaerobic Wastewater Treatment." *Journal of the Chinese Institute of Chemical Engineers* 39(6):557–562.

Spilker, Gregor, and Nick Nugent. 2022. "Voluntary Carbon Market Derivatives: Growth, Innovation & Usage." *Borsa Istanbul Review* 22:S109–S118.

Swan, Melanie. 2015. *Blockchain: Blueprint for a New Economy.* O'Reilly Media, Inc.

Teufel, Bernd, Anton Sentic, and Mathias Barmet. 2019. "Blockchain Energy: Blockchain in Future Energy Systems." *Journal of Electronic Science and Technology* 17(4):100011.

Vargas, Don Tapscott, and Ricardo Viana. 2019. "How Blockchain Will Change Construction." *Harvard Business Review.*

Vilkov, Arsenii, and Gang Tian. 2023. "Blockchain's Scope and Purpose in Carbon Markets: A Systematic Literature Review." *Sustainability* 15(11):8495.

Wang, Qiang, and Min Su. 2020. "Integrating Blockchain Technology into the Energy Sector – from Theory of Blockchain to Research and Application of Energy Blockchain." *Computer Science Review* 37:100275.

Weikmans, Romain, Harro van Asselt, and J. Timmons Roberts. 2021. "Transparency Requirements under the Paris Agreement and Their (Un)Likely Impact on Strengthening the Ambition of Nationally Determined Contributions (NDCs)." Pp. 107–122 in *Making Climate Action More Effective.* Routledge.

Woo, Junghoon, Ashish T. Asutosh, Jiaxuan Li, Wolfgang D. Ryor, Charles J. Kibert, and Alireza Shojaei. 2020. "Blockchain: A Theoretical Framework for Better Application of Carbon Credit Acquisition to the Building Sector." Pp. 885–894 in *Construction Research Congress 2020*. American Society of Civil Engineers.

Woo, Junghoon, Ridah Fatima, Charles J. Kibert, Richard E. Newman, Yifeng Tian, and Ravi S. Srinivasan. 2021. "Applying Blockchain Technology for Building Energy Performance Measurement, Reporting, and Verification (MRV) and the Carbon Credit Market: A Review of the Literature." *Building and Environment* 205:108199.

Yang, Zhihan, Chen Zhu, Yimin Zhu, and Xiaodong Li. 2023. "Blockchain Technology in Building Environmental Sustainability: A Systematic Literature Review and Future Perspectives." *Building and Environment* 245:110970.

Yuan, Xin, Shusheng Qian, and Bing Li. 2023. "Carbon Neutrality: A Review." *Journal of Computing and Information Science in Engineering* 23:60801–60809.

Zaid, Mohamed, and Noor Suzaini. 2013. "Measuring Electricity-Related GHG Emissions and the Affordability of Electricity in Malaysian Low-Cost Housing: A Case Study of Low-Cost Housing Projects in Kuala Lumpur." UNSW Sydney.

Zhang, L. P., and P. Zhou. 2018. "A Non-Compensatory Composite Indicator Approach to Assessing Low-Carbon Performance." *European Journal of Operational Research* 270(1):352–361.

Zhang, Xiaofang. 2020. "Emissions Trading Scheme in Japan: Based on Tokyo and Saitama Cases." 千葉大学人文公共学研究論集 = *Journal of Studies on Humanities and Public Affairs of Chiba University* 41:143–150. https://cir.nii.ac.jp/crid/1050851497148557312.

Index

For Product Safety Concerns and Information please contact our EU
representative GPSR@taylorandfrancis.com
Taylor & Francis Verlag GmbH, Kaufingerstraße 24, 80331 München, Germany

www.ingramcontent.com/pod-product-compliance
Ingram Content Group UK Ltd.
Pitfield, Milton Keynes, MK11 3LW, UK
UKHW022313100726
473146UK00009B/460